A Study Guide and Map Supplement

for

The Western Experience
Volume II
Since the Sixteenth Century

Eighth Edition

A Study Guide and Map Supplement

for

The Western Experience
Volume II
Since the Sixteenth Century

Eighth Edition

Mortimer Chambers
University of California, Los Angeles

Raymond Grew
University of Michigan

Barbara Hanawalt
University of Minnesota

Theodore K. Rabb
Princeton University

Isser Woloch
Columbia University

Prepared by

Edward W. M. Bever
State University of New York at Old Westbury

Boston   Burr Ridge, IL   Dubuque, IA   Madison, WI   New York   San Francisco   St. Louis
Bangkok   Bogotá   Caracas   Kuala Lumpur   Lisbon   London   Madrid   Mexico City
Milan   Montreal   New Delhi   Santiago   Seoul   Singapore   Sydney   Taipei   Toronto

A Study Guide and Map Supplement for
THE WESTERN EXPERIENCE: VOLUME II: SINCE THE SIXTEENTH CENTURY
Chambers, Grew, Hanawalt, Rabb, Woloch

Published by McGraw-Hill, an imprint of The McGraw-Hill Companies, Inc., 1221 Avenue of the Americas, New York, NY 10020. Copyright © 2003, 1999, 1995, 1991 by The McGraw-Hill Companies, Inc. All rights reserved. Previous editions © 1987, 1983, 1981, 1979 by Alfred A. Knopf, Inc. All rights reserved. No part of this publication may be reproduced or distributed in any form or by any means, or stored in a database or retrieval system, without the prior written consent of The McGraw-Hill Companies, Inc., including, but not limited to, in any network or other electronic storage or transmission, or broadcast for distance learning.

1 2 3 4 5 6 7 8 9 0 BKM/BKM 0 9 8 7 6 5 4 3 2 1

ISBN 0-07-256557-8

www.mhhe.com

ABOUT THE AUTHORS

Dennis Sherman is Professor of History at John Jay College of Criminal Justice, the City University of New York. He received his B.A. (1962) and J.D. (1965) from the University of California at Berkeley and his Ph.D. (1970) from the University of Michigan. He was Visiting Professor at the University of Paris (1978–1979; 1985). He has received the Ford Foundation Prize Fellowship (1968–1969, 1969–1970), a fellowship from the Council for Research on Economic History (1971–1972), and fellowships from the National Endowment for the Humanities (1973–1976). His publications include *A Short History of Western Civilization,* Eighth Edition, 1994 (co-author); *Western Civilization: Sources, Images, and Interpretations,* Fourth Edition, 1995; *World Civilizations: Sources, Images, and Interpretations,* 1994 (co-author); a series of introductions in the Garland Library of War and Peace; several articles and reviews on nineteenth-century French economic and social history in American and European journals; and short fiction in literary journals.

Edward Bever is Assistant Professor of History at the State University of New York College at Old Westbury. He received his B.A. (1975) from Dartmouth College and his M.A. (1978) and Ph.D. (1983) from Princeton University. From 1984 to 1997 he worked as a software developer specializing in historical simulations and reference works. During this time he taught a variety of courses in European and African history and Western Civilization at Drew University, Purchase College, Loyola College of Maryland, Monmouth College, and Mercer County College in New Jersey. His software creations include *Crusade in Europe* (1985), *Conflict in Vietnam* (1986), *Revolution '76* (1989) and *No Greater Glory* (1991), and he wrote a series of essays and designed a set of maps surveying Western Civilization for the multimedia reference work *Culture 2.0* (1991). His written publications include a book on African political history, *Africa* (1996); extensive historical background material to accompany his software; and several scholarly articles, reviews, and papers on popular culture and witchcraft trials in early modern Europe.

CONTENTS

Preface vii
Introduction viii

15 War and Crisis 1
16 Culture and Society in the Age of the Scientific Revolution 12
17 The Emergence of the European State System 22
18 The Wealth of Nations 33
19 The Age of Enlightenment 42
Section Summary The Early Modern Period 1560–1789 50

20 The French Revolution 55
21 The Age of Napoleon 68
22 Foundations of the Nineteenth Century: Politics and Social Change 78
23 Learning to Live with Change 88
24 National States and National Cultures 98
25 European Dynamism and the Nineteenth-Century World 111
26 The Age of Progress 122
Section Summary The Nineteenth Century 1789–1914 130

27 World War and Democracy 137
28 The Great Twentieth-Century Crisis 149
29 The Nightmare: World War II 161
30 Contemporary Europe 174
Section Summary The Twentieth Century 1914–Present 185

Epilogue 192

PREFACE

This new edition of the *Study Guide* reflects changes in the Eighth Edition of *The Western Experience* while maintaining the enhanced presentation of material introduced in the Seventh Edition *Guide* that made it more interactive and more self contained.

Specifically, the enhanced presentation includes sets of multiple choice, self test questions in place of the older chapter summaries; the Guide to the Documents sections have some multiple choice questions as well; and the Significant Individuals and Identification sections are structured as matching exercises. These formats make using the *Guide* more interactive, while inclusion of the answers at the end of each chapter makes it possible to get immediate feedback about how well material in the textbook is being understood. The Chapter Highlights, Chapter Outline, and Transitions sections still summarize the main points of the chapters, and the Problems for Analysis, Speculations, and some of the Guide to the Documents questions are presented in the open-ended, "essay question" style that is generally stressed in college courses, so the overall effect is to combine the values of the new and old approaches to the material.

The section summaries keep both of the features introduced in the Seventh Edition. The first of these is the Cultural Styles sections, which pose questions about works of art from different civilizations and eras in order to focus attention on the visual information contained in the textbook's rich set of illustrations about the characteristics of, links between, and differences among the periods and traditions of Western Civilization. The second is the structuring of the Box Charts so that they provide a framework in which to organize the material according to the overarching themes identified by the authors in the textbook. This feature is even more important in this new edition because it supports the authors' increased emphasis on these structures in the body of the text. Using these charts, the long-term development of Western Civilization can be followed within the authors' standard frame of reference, helping to bring order and meaning to the immense and complex story of *The Western Experience*.

INTRODUCTION

WHAT IS HISTORY?

Definition

History is the record of the human past. It includes both the more concrete elements of the past such as our wars, governments, and creations, and the more elusive ones such as hopes, fantasies, and failures over time. Historians study this human past in order to discover what people thought and did, and then they organize these findings into a broad chronological framework. To do this, they look at the records humans have left of their past, the most important of which are written. Although nonwritten records, such as artifacts, buildings, oral traditions, and paintings, are also sources for the study of the human past, the period before written records appear is usually considered "pre-history."

The Purposes of History

History can be used for various purposes. First, a systematic study of the past helps us to understand human nature; in short, history can be used to give us an idea of who we are as human beings. Second, it can be used to gain insights into contemporary affairs, either through a study of the developments that have shaped the present or through the use of analogies to related circumstances of the past. Third, societies use history to socialize the young, that is, to teach them how to behave and think in culturally and socially appropriate ways.

Orientation

Historians approach the study of history from two main perspectives: the humanities and the social sciences. Those with a humanities orientation see history as being made up of unique people, actions, and events, which are to be studied both for their intrinsic value and for the insights they provide about humans in a particular set of historical circumstances. Those with a social science orientation look for patterns in human thought and behavior over time. They focus on comparisons rather than on unique events and are more willing to draw conclusions related to present problems. The authors of *The Western Experience,* Eighth Edition, use both perspectives.

Styles

In writing history, historians traditionally use two main styles: narrative and analytic. Those who prefer the narrative style emphasize a chronological sequence of events. Their histories are more like stories, describing the events from the beginning to the end. Historians who prefer the analytic style emphasize explanation. Their histories deal more with topics, focusing on causes and relationships. Most historians use both styles but show a definite preference for one or the other. The authors of *The Western Experience* stress the analytic style.

Interpretations

Some historians have a particular interpretation or philosophy of history, that is, a way of understanding its meaning and of interpreting its most important aspects. Marxist historians, for example, argue that economic forces are most important, influencing politics, culture, and society in profound ways. They view history organically, following a path in relatively predictable ways. Their interpretations are occasionally pointed out in *The Western Experience.* Most historians are not committed to a particular philosophy of history, but they do interpret major historical developments in certain ways. Thus, for example, there are various groups of historians who emphasize a social interpretation of the French

Revolution of 1789, while there are others who argue that the Revolution is best understood in political and economic terms. The authors of *The Western Experience* are relatively eclectic; they use a variety of interpretations and often indicate major points of interpretive disagreement among historians.

Common Concerns

The style, orientation, and philosophy of most modern historians are not at one extreme or the other. Moreover, the concerns they share outweigh their differences. All historians want to know what happened, when it happened, and how it happened. While the question of cause is touchy, historians all want to know why something happened, and they are all particularly interested in studying change over time.

THE HISTORICAL METHOD

Process

    1. *Search for Sources:* One of the first tasks a historian faces is the search for sources. Most sources are written documents, which include everything from gravestone inscriptions and diaries to books and governmental records. Other sources include buildings, art, maps, pottery, and oral traditions. In searching for sources, historians do not work at random. They usually have something in mind before they start, and in the process, they must decide which sources to emphasize over others.

    2. *External Criticism:* To test the genuineness of the source, historians must engage in external criticism. This constitutes an attempt to uncover forgeries and errors. Some startling revisions of history have resulted from effective criticism of previously accepted sources.

    3. *Internal Criticism:* A source, though genuine, may not be objective, or it may reveal something that was not apparent at first. To deal with this, historians subject sources to internal criticism by such methods as evaluating the motives of the person who wrote the document, looking for inconsistencies within the source, and comparing different meanings of a word or phrase used in the source.

    4. *Synthesis:* Finally, the historian creates a synthesis. He or she gathers the relevant sources together, applies them to the question being investigated, decides what is to be included, and writes a history. This oversimplifies the process, for historians often search, criticize, and synthesize at the same time. Moreover, the process is not as objective as it seems, for historians select what they think is most important and what fits into their own philosophy or interpretation of history.

Categories

Historians use certain categories to organize different types of information. The number and boundaries of these differ according to what each historian thinks is most useful. The principal categories are as follows:

    1. *Political:* This refers to questions of how humans are governed, including such matters as the exercise of power in peace and war, the use of law, the formation of governments, the collection of taxes, and the establishment of public services.

    2. *Economic:* This refers to the production and distribution of goods and services. On the production side, historians usually focus on agriculture, commerce, manufacturing, and finance. On the distribution side, they deal with who gets how much of what is produced. Their problem is supply and demand and how people earn their living.

3. *Social:* This is the broadest category. It refers to relations between individuals or groups within some sort of community. This includes the institutions people create (the family, the army), the classes or castes to which people belong (the working class, the aristocracy), the customs people follow (marriage, eating), the activities people engage in together (sports, drinking), and the attitudes people share (toward foreigners, commerce).

4. *Intellectual:* This refers to the ideas, theories, and beliefs expressed by people in some organized way about topics thought to be important. This includes such matters as political theories, scientific ideas, and philosophies of life.

5. *Religious:* This refers to theories, beliefs, and practices related to the supernatural or the unknown. This includes such matters as the growth of religious institutions, the formation of beliefs about the relation between human beings and God, and the practice of rituals and festivals.

6. *Cultural:* This refers to the ideas, values, and expressions of human beings as evidenced in aesthetic works, such as music, art, and literature.

While most historians work with these categories on a relatively *ad hoc* basis, some attempt to define them in a way that provides a comprehensive structure to history and human experience. The authors of *The Western Experience* discuss in the introduction to the textbook the categories that they see as structuring human affairs: Social Structures (Groups in Society), Political Events and Structures, the Economics of Production and Distribution, Family (including Gender Roles and Daily Life), War, Religion, and Cultural Expression (including intellectual life). While they do not use these categories explicitly to structure the book, they incorporate them implicitly throughout it, and there is a chart at the beginning of each chapter showing the categories the chapter focuses on.

In addition to organizing different types of information into categories, historians often specialize in one or two of these. For example, some historians focus on political history, whereas others are concerned with social-economic history. The best historians bring to bear on the problems that interest them, however specialized the problems may seem, data from all these categories.

DOCUMENTS

Historians classify written documents into two types: primary and secondary. Primary documents are those written by a person living during the period being studied and participating in the matter under investigation. A primary document is looked at as a piece of evidence that shows what people thought, how they acted, and what they accomplished. A secondary document is usually written by someone after the period of time that is being studied. It is either mainly a description or an interpretation of the topic being studied: the more descriptive it is, the more it simply traces what happened; the more interpretive it is, the more it analyzes the causes or the significance of what happened.

PERIODIZATION

Historians cannot deal with all of history at once. One way to solve this is to break history up into separate periods. How this is done is a matter of discretion; what is important is the division of a time into periods that can be dealt with as a whole, without doing too much violence to the continuity of history. Typically, Western Civilization is divided into the Ancient World, the Middle Ages, the Renaissance and Reformation, the Early Modern World, the nineteenth century (1789–1914), and the twentieth century (1914–present), as illustrated by the section summaries in this study guide. There are a number of subdivisions that can be made within these periods. *The Western Experience* is divided into both periodic and topical chapters.

STUDY AIDS

Reading

There is no way to get around reading—the more you read, the better you become at it. Thus the best advice is to read the assigned chapters. But there are some techniques that will make the task easier. First, think about the title of the book you are reading; often, it tells much about what is in the book. Second, read through the chapter headings and subheadings. Whatever you read will make more sense and will be more easily remembered if it is placed in the context of the section, the chapter, and the book as a whole. Third, concentrate on the first and last paragraphs or two of each chapter and each major section of the chapter. Often, the author will summarize in these areas what he or she wants to communicate. Finally, concentrate on the first sentence of each paragraph. Often, but not always, the first sentence is a topic sentence, making the point for which the rest of the paragraph is an expansion.

Underlining (or Highlighting) and Writing in the Margins

The easiest and quickest way to begin mastering the material in the book is to underline or highlight important points and ideas as you go along. An underlined or highlighted passage can be surveyed and reviewed much more quickly than a clear one, whether the purpose is to take notes (see below), to find material for an essay, or to review for a test. Comments in the margins can also be useful; they can be made to summarize long passages, to call out key terms, or to remind yourself about your reactions to the author's points.

Some students are reluctant to mark in books, either out of a misguided reverence for printed material or out of an equally misguided desire to resell the text back to the bookstore at the end of the course. The latter concern is simply short-sighted, for considering the cost of a college course, the cost of the textbook is minor, and avoiding techniques that will help wring the full value from the major investment to recoup the minor one is being penny wise and pound foolish. The former concern, a reverential attitude towards printed materials, is a vestige from long ago, when books were rare and precious things. They are still precious, but they are not rare, and there's nothing wrong with a well-used textbook looking well-used. You own the book; make it your own.

Note that this approach should NOT be taken to library books. Other people need to use them for their own purposes, so personalizing them in this way is inconsiderate.

Note Taking

   1. *Reading Assignments:* While it is easier said than done, it is of tremendous advantage to take notes on reading assignments. Taking notes, if done properly, will help you to integrate the readings into your mind much more than simply reading them. Moreover, notes will ease the problem of review for papers, exams, or classroom discussions.

   There are a number of ways to take notes. Generally, you should use an outline form, following the main points or headings of each chapter. Under each section of your outline you should include the important points and information, translated into your own words. After each section of a chapter ask yourself, "What is the author trying to say here, what is the author trying to convey?" It may be easier to copy phrases or words used by the author, but it is much more effective if you can transform them into your own words. While facts, names, and dates are important, avoid simply making a list of them without focusing on the more general interpretation, development, or topic that the author is discussing. Indicate the kinds of evidence the author uses, what the author's interpretations are, and to what degree you agree with what he or she says (does it make sense to you?). Some students prefer to underline in the text and write notes in the margins. This is a less time-consuming, easier, and often useful method, but probably not as effective as outlining and using your own words to summarize each section. As with many things, it is the extra effort involved that leads to the more effective learning.

2. *Lectures and Classes:* Much of what has been said also applies to taking notes in class. The trick is to write just enough to get the main points without losing track of what is going on in class. Concentrate on the major points the speaker is trying to make, not simply all the facts. Do not try to write down everything or to write in complete sentences; try to develop a method of writing down key words or phrases that works for you. The most important thing is the act of taking notes, for taking notes involves you actively with the class material. Not taking notes, in contrast, leads to a passive state of listening, and soon, daydreaming, at which point much of what is said will go in one ear and out the other.

A couple of additional tips might help. First, be ready at the beginning of class. Often, the point of a whole lecture or discussion is outlined in the first couple of minutes; missing it makes much of what you hear seem out of context. Second, go over your notes after class. A few minutes spent reviewing the main points while they are fresh in your mind can make studying the material later much easier.

Writing

There are three steps that you should take before writing a paper. First, carefully read the question or topic you are to write on; at times, good papers are written on the wrong topic. If you are to make up a topic, spend some time on it. Think of your topic as a question. It should not be unanswerably broad (what is the history of Western Civilization?) or insignificantly narrow (when was toothpaste invented?). It will be something that interests you and that is easily researched. Second, start reading about the topic, taking notes on the main points. Third, after some reading, start writing an outline of the main points you want to make. Revise this outline a number of times, arranging your points in some logical way and making sure all your points help answer the question or support the argument you are making.

A paper should have three parts: an introduction, a body, and a conclusion. For a short paper, your introduction should be only a paragraph or two in length; for a long paper, perhaps one or two pages. In the introduction tell the reader what the general topic is, what you will argue about the topic, and why it is important or interesting. This is an extremely important part of a paper, often neglected by students. You can win or lose the reader with the introduction. You may find it easier to leave the introduction until after you have written the body of the paper, especially if it is difficult to get started writing. In the body, make your argument. Generally, make one major point in each paragraph, usually in the first sentence (topic sentence). The rest of the paragraph should contain explanation, expansion, support, illustration, or evidence for this point. In the paragraph, you should make it clear how this point helps answer the question. Your paragraphs should be organized in some logical order (chronological, from strongest to weakest point, categories). Finally, in the conclusion, tell the reader—in different words—what you have argued in the body of the paper, and indicate why what you have argued is important. The conclusion, like the introduction, is a particularly important and yet an often slighted part of a paper.

Most of the same suggestions for papers apply to essay exams. Even more emphasis should be placed on making sure you know what the question asks. Spend some time outlining your answer. As with papers, you should have an introduction, body, and conclusion, even if they are all relatively short. For each point you make, try to supply some evidence as support. Keep to the indicated time limits.

Class Participation

Class participation is difficult for many students, yet there is no better way to get over this difficulty than to do it. Try and force yourself to ask questions or indicate your point of view when appropriate times arise. If this is particularly difficult for you, it may help to talk about it with other students or with the teacher privately.

Studying

If you have a style of studying that works well for you, stick to it. If not, try to do three things: keep up with all your assignments regularly, work with someone else, and spend some extra time reviewing before exams.

HOW TO USE THIS GUIDE

There is one chapter in this guide corresponding to each chapter in *The Western Experience*. Each chapter of the guide is divided into a number of sections.

1. *Chapter Highlights:* The main points of each chapter are introduced here. Part of the purpose in doing this is to emphasize the importance of not losing sight of the broader concerns of the chapter as you study its specific sections. By returning to these main themes and expanding upon them after you read the chapter, they can become a tool to help you grasp more firmly what the chapter is about.

2. *Chapter Outline:* Here, the chapter is outlined using the major headings in the text. The purpose is to provide an overview of the chapter so that what you read will make more sense and will be more easily remembered because it can be seen in the context of the section, the chapter, and the book as a whole.

3. *Self Test:* This section contains approximately 20 multiple choice questions. In general, there are one or two questions per sub-heading, and they are presented in the order of the sections in the text. The purpose of these questions is to help focus your reading of the text. Note that they are not written as mock-test questions, but instead are styled to help you master the material. Thus, there are many questions that ask which one of the four possibilities is NOT correct, so that you will spend more time thinking about the correct information than about wrong answers. Many of the questions have been written to emphasize the broad concepts or range of information that are important for you to know. Others have been made deliberately difficult, turning on very specific details, in order to draw you back to the book. You can use the section as a simple review test to check your knowledge, but you will get more out of it by using it as a more active study guide. Pay as much attention to the answers that you don't select as the ones that you do, plan on going back to the book to find the answers, and don't stop yourself from re-reading other things that draw your attention while you're there.

4. *Guide to Documents:* Each chapter contains questions related to the documents used in *The Western Experience*. In general, the first question is a multiple choice question focusing on your comprehension of the main point of the document. The second is then an "essay" style question calling for broader interpretive thought and showing how the document might be used to increase historical understanding and gain insights into historical questions.

5. *Significant Individuals:* Here, the principal historical figures mentioned in the text are listed, along with some brief biographical information. This is intended to be used in two ways: first, as an exercise; and second, as a reference. The section is structured as two columns, with the names in a numbered list on the left, and the descriptions in a lettered list on the right. Note that the names are NOT next to the correct description. The first thing you need to do is the exercise of matching the name to the correct description. You can do this while you are reading or afterwards, and either by drawing a line from the name to the description or by writing the description's letter next to the name. Once all the names and descriptions have been matched, you can use the list as a reference when you study.

6. *Chronological Diagram:* This diagram is intended to be used as a reference. Note what different sorts of events are related chronologically. It is often useful to compare the chronological chart in one chapter with those in the preceding and succeeding chapters. On an even broader scale, this is done in the chronological diagrams contained in each section summary.

7. *Identification:* Some of the most important developments and events in the chapter are listed here. It is set up exactly like the Significant Individuals section, and you can use it in the same ways.

8. *Map Exercises:* In most chapters maps are provided with exercises that relate to some of the main concerns of the chapter. Standing alone and without directly using the text, some of these exercises are difficult. But by utilizing the maps already present in the text and in some cases specific sections of the text referred to in the exercise, they become easier. The purposes here are to help you get used to using maps, to emphasize the importance of geographic considerations in history, and to encourage you to picture developments described in the text in concrete, geographic terms.

9. *Problems for Analysis:* These are designed to cover each of the main sections of the chapter. They require a combination of specific information and analysis. Working on these problems should give you a much stronger grasp of the materials and issues dealt with in each section of the chapter. In addition, you might use some of these problems to prepare for class discussions. They might help you formulate questions to ask in class or present a point of view that you find particularly interesting or irritating.

10. *Speculations:* These constitute unusual, interesting questions. They may require you to put yourself back into history, compare the past with the present, or speculate on various historical alternatives. They might be used as a first step toward identifying a paper topic or developing a classroom debate. From aspects of broader speculations, more specific historical problems could be identified, put into perspective, and dealt with.

11. *Transitions:* These relate the previous chapter, the present chapter, and the following chapter. One of the main purposes here is to help you avoid losing the continuity of history; each chapter in the text is integrally connected to what came before and what follows. Another purpose is to emphasize briefly the main arguments presented in the chapter; focusing on specific events can sometimes lead one to overlook the broader conclusions that are being drawn.

12. *Answers:* The answers to all multiple choice questions are given at the end of each chapter. You can use these answers in three ways. (1) If you want, you can simply come to this section, mark the answers near the questions, and use the exercises as a reference. (2) You can answer the questions from memory to test yourself, and then come here to see how well you did. Study the sections were you get question wrong particularly carefully. (3) You can avoid this section as much as possible by referring back to the book to check your answers or help find the answers you know you don't know. Only check here to make sure you have come up with the right answer on your own. This is probably the most effective way to use the multiple choice questions and matching exercises as an aid to your studies.

In addition to the chapters, there are three section summaries. These correspond to periods into which historians commonly divide Western history and to sections of the book often covered in an exam or a paper. Each contains Chronological Diagrams, Map Exercises, and Box Charts for you to fill in. Tabulating material from your reading notes on these charts (which you will need to reproduce in larger format in your notebook or on separate sheets of paper) will help you place individuals and events in a broader thematic and chronological framework and distinguish important facts from less important ones. Note that the charts are structured according to the categories discussed above in order to give you a standard frame of reference into which to fit the history of Western Civilization. The authors discuss these categories in the introduction to the textbook, include a chart at the beginning of each chapter identifying which themes the chapter focuses on, and have organized the index to make it relatively easy to access material on each topic. The charts should be particularly useful when you are reviewing for an exam.

Dennis Sherman
Edward Bever

FIFTEEN
WAR AND CRISIS

CHAPTER HIGHLIGHTS

1. A series of costly, devastating wars, inflamed by religious motives, raged in Europe in the period between the 1560s and 1650s.
2. The most devastating war was the Thirty Years' War. It was brought to an end by the Peace of Westphalia, which signified major international changes and a new period of relative calm.
3. New weapons, tactics, and armies revolutionized war during the period.
4. Throughout Europe tensions created great internal unrest, often breaking out in revolt and civil war.
5. During the mid-seventeenth century, new political patterns were established that would hold for some time.

CHAPTER OUTLINE

I. Rivalry and War in the Age of Philip II
1. Philip II
2. Elizabeth I of England
3. The Dutch Revolt
4. Civil War in France

II. From Unbounded War to International Crises
1. The Thirty Years' War
2. The Peace of Westphalia

III. The Military Revolution
1. Weapons and Tactics
2. The Organization and Support of Armies
3. The Life of the Soldier

IV. Revolution in England
1. Pressures for Change
2. Parliament and the Law
3. Rising Antagonisms
4. Civil War
5. England under Cromwell

V. Revolts in France and Spain
1. The France of Henry IV
2. Louis XIII
3. Political and Social Crisis
4. The Fronde
5. Sources of Discontent in Spain
6. Revolt and Secession

VI. Political Change in an Age of Crisis
1. The United Provinces
2. Sweden
3. Eastern Europe and the Crisis

SELF TEST

1. Philip II of Spain dominated the second half of the sixteenth century because of his
    a. insatiable quest for military glory.
    b. determination to defeat the enemies of Catholicism.
    c. goal of securing Spanish dominance of Europe.
    d. desire to restore the empire of his father, Charles V.

2. England's victory over the Spanish Armada accomplished all of the following EXCEPT
    a. preventing an invasion of England by Spanish troops.
    b. enabling England to continue supporting the Dutch rebels.
    c. sparking rebellions in Portugal, Catalonia, Naples, and Sicily.
    d. sealing the fate of Spain's Catholic allies in the French Civil War.

3. The Dutch revolt was all of the following EXCEPT
    a. a national struggle against a foreign overlord.
    b. a religious struggle between Protestants and Catholics.
    c. the first major victory in Europe by subjects resisting their monarch's authority.
    d. the first of the religious wars to end with a treaty ensuring tolerance for both confessions.

4. All of the following were issues in the French Civil War EXCEPT
    a. Philip II's desire to draw France into his dynastic empire.
    b. the struggle between the Protestants and the Catholics.
    c. the rivalry of the Guise and the Bourbon families.
    d. the reassertion of autonomy by the great nobles.

5. All of the following were true of the combatants in the Thirty Years' War EXCEPT
    a. The Habsburgs sought to defeat Protestantism and establish control over the Holy Roman Empire.
    b. The German Catholics sought to advance their religion while avoiding Habsburg domination.
    c. The German Protestants sought to defend their religion and avoid Habsburg domination.
    d. The Swedes, Spanish, and French sought to keep the Empire weak to enhance their own relative strength.

6. The Peace of Westphalia was important for all of the following reasons EXCEPT
    a. it ended the anarchy in Germany.
    b. it secured Spanish control of Holland.
    c. it formalized the fragmentation of Germany.
    d. it laid the groundwork for international relations for the next century.

7. The primary cause of the military revolution was
    a. gunpowder.
    b. pikemen.
    c. sieges.
    d. discipline.

8. Between 1550 and 1700, the size of the leading army in Europe increased
    a. from 40,000 to 60,000 men.
    b. from 40,000 to 100,000 men.
    c. from 40,000 to 200,000 men.
    d. from 40,000 to 400,000 men.

9. The growth of the military caused all of the following to increase EXCEPT
    a. the size of the government bureaucracy needed to support the military.
    b. the amount of taxes needed to support the military.
    c. the amount of damage done by soldiers to the areas they were stationed.
    d. the elaboration of military command and administrative structures.

10. All of the following social groups were centers of opposition to the English monarch EXCEPT
    a. the Puritans.
    b. the gentry.
    c. the nobility.
    d. the merchants.

11. The Petition of Rights called for and end to all of the following EXCEPT
    a. imprisonment without cause shown.
    b. the king's right to dissolve Parliament.
    c. taxation without Parliament's consent.
    d. martial law in peacetime.

12. The major factions in the English Civil War included all of the following EXCEPT
    a. Royalists, who supported the king once the Grand Remonstrance was passed.
    b. Presbyterians, Puritans who wanted a centrally organized, Calvinist Church.
    c. Lutherans, who wanted to make the original Protestantism as the state religion.
    d. Independents, Puritans who wanted each congregation to rule itself.

13. The English Civil War ended with the Restoration of the Stuart dynasty because
    a. the people had never supported revolution and turned on the Puritans once Cromwell was dead.
    b. the rebels were unable to create viable permanent structures.
    c. Parliament realized that it had acted illegally and wanted to restore the rule of law.
    d. royalists were able to infiltrate the Parliamentary government and stage a bloodless coup.

14. Henry IV's accomplishments included all of the following EXCEPT
    a. creating religious peace by converting to Catholicism while guaranteeing the rights of the Protestants.
    b. reestablishing the authority of the king by buying off the nobility and the principal bureaucrats.
    c. establishing the notion that the government had primary responsibility to foster economic development.
    d. leading the French intervention in the Holy Roman Empire that frustrated Habsburg designs.

15. Cardinal Richelieu accomplished all of the following EXCEPT
    a. reducing the independence of the traditional nobles.
    b. establishing the *intendents* as dominant officials in the provinces.
    c. destroying the Huguenots' independent military and political power.
    d. leading the French to final victory in the Thirty Years' War.

16. The growth of royal power in France caused discontent in all of the following groups EXCEPT
    a. peasants, who objected to the increasingly onerous taxes.
    b. officials, who wanted to retain and expand their traditional prerogatives.
    c. merchants, who opposed the heavy hand of government regulation.
    d. nobles, who resisted the crowns' reduction of their traditional autonomy.

17. The Fronde failed because of
    a. Spanish intervention.
    b. the disunity of the rebels.
    c. the perfidy of the nobles.
    d. the prestige of the king.

18. Spain suffered from all of the following problems in the mid-seventeenth century EXCEPT
    a. the wealth from America had been squandered on wars rather than invested in economic development.
    b. the bureaucracy was inefficient and dominated by Castilians, who were resented in other provinces.
    c. devastating plagues reduced the population by 40%, from 10 million to 6 million people.
    d. foreign enemies had invaded the country and seized its most valuable territories.

19. Which of the following was able to break away from the Spanish control during the revolts at mid-century?
    a. Portugal.
    b. Catalonia.
    c. Sicily.
    d. Naples.

20. The primary constitutional struggle in the United Provinces was between
    a. the House of Orange, backed by the rural provinces, and the merchant oligarchy that controlled Holland.
    b. the House of Orange, backed by the merchant oligarchy in Holland, and the rural provinces.
    c. the Protestant northern provinces and the Catholic southern ones.
    d. the Protestant urban provinces and the Catholic rural ones.

21. Gustavus Adolphus accomplished all of the following for Sweden EXCEPT
    a. he made it the most powerful state in the Baltic area and a major power in Europe.
    b. he organized a bureaucracy that was superior to most until the twentieth century.
    c. he tamed the nobility so the country was spared constitutional turmoil for centuries.
    d. he fostered the development of the country's mines and other economic assets.

22. The Romanov dynasty in Russia accomplished all of the following EXCEPT
    a. cementing an alliance with the nobility.
    b. forcing the peasantry into complete subordination.
    c. codifying the laws and establishing control of the church.
    d. keeping the Ukraine from switching allegiance to Poland.

GUIDE TO DOCUMENTS

I. Queen Elizabeth's Armada Speech

1. How did Elizabeth attempt to appeal to the troops?
    a. She emphasized her trust in and commitment to them.
    b. She played on their chivalric duty to protect her as a woman.
    c. She emphasized their national superiority over the Spaniards.
    d. She whipped up their fear and hatred of Catholics.

2. In what ways does she justify her own political power?

II. Oliver Cromwell's Aims

1. Cromwell gives all of the following reasons for declining the offer to serve as king EXCEPT
    a. he wants Parliament to create the long term basis for peace while he maintains order in the short term.
    b. he thinks that as Constable he already has sufficient power to settle English peace and liberties.
    c. God has destroyed the office as well as the person of the king, and he will not go against His Providence.
    d. the destruction of the monarchy was decided by a prolonged and convulsive process.

2. How does Cromwell's reasoning reflect his religious beliefs and the role religion played in the civil war?

III. Richelieu on Diplomacy

1. All of the following insights conveyed in this excerpt help explain Richelieu's extraordinary success EXCEPT
   a. he pursued every avenue of negotiation that might lead to success, and many did.
   b. he pressed ahead with critical negotiations to maximize the benefit gained from them.
   c. he used his contacts to gain as much information as possible even when the negotiations led nowhere.
   d. he avoided conceding anything of value in negotiations, calculating that eventually he would get his way.

2. In what ways does the document reflect the connections between diplomacy and international interests of the state during this period?

SIGNIFICANT INDIVIDUALS

1. Philip II
2. Elizabeth I
3. William of Orange
4. Maurice of Nassau
5. Catherine de Medici (MED-a-chē)
6. Duke of Guise (gēz)
7. Henry IV
8. Ferdinand II
9. Albrecht von Wallenstein (WOL-en-stīn)
10. Gustavus Adolphus (gus-TAH-vus a-DOL-fus)
11. James I
12. Charles I
13. Oliver Cromwell
14. Louis XIII
15. Cardinal Richelieu (rē-shuh-LYOO)
16. Cardinal Mazarin (MAZ-e-rin)
17. Count of Olivares (Ō-lē-var-ez)
18. Jan De Witt (yahn de wit)

a. Mercenary general in Thirty Years' War (1583-1634)
b. Spanish minister who sparked revolts (r.1621-1643)
c. French minister who built up monarchy (r.1624-1642)
d. Scottish king who became King of England and began struggle with Parliament (r.1603-1625)
e. Swedish king and great general (r.1611-1632)
f. Brilliant military leader of Dutch revolt (1587-1625)
g. French minister who defeated Fronde (r.1642-1661)
h. Independent, general, and ruler of England (1599-1658)
i. Leader of French Catholics in Wars of Religion (1550-1589)
j. English queen who defeated Spanish armada (r.1558-1603)
k. Representative of mercantile Holland (1625-1672)
l. French king during Thirty Years War (r.1610-1643)
m. Initial leader of Dutch revolt (1533-1584)
n. English king who lost his head (r.1625-1649)
o. Regent during French Wars of Religion (1519-1589)
p. Emperor who provoked Thirty Years War (r.1619-1637)
q. French king who ended Wars of Religion (r.1589-1610)
r. King of Spain at its height of power (r.1556-1598)

IDENTIFICATION

1. Spanish Armada
2. Huguenots (HYOO-ge-nots)
3. the Catholic League
4. *politiques* (pol-i-TIKS)
5. Edict of Nantes (nahnts)
6. Peace of Westphalia (west-FĀ-lē-a)
7. the salvo
8. Grand Remonstrance
9. "Long" Parliament
10. New Model Army
11. "Rump" Parliament
12. Levellers
13. *paulette* (pol-et)
14. mercantilism
15. *intendant* (in-TEN-dant)
16. the Fronde (frond)
17. Union of Arms
18. "general crisis"

a. Treaty that framed European diplomatic order after 1648
b. Fleet that tried to invade England
c. Series of revolts in mid-seventeenth century France
d. Tactic of having all musketeers fire at once
e. Representative body that sat through English Civil War
f. Force commanded by Cromwell
g. French faction in favor of stability through monarchy
h. Doctrine that government should build up economy
i. Military organization of French Catholics
j. English social revolutionaries
k. French royal official used to reduce local nobles' power
l. French royal decree setting limited religious toleration
m. Sum of legislation passed at start of English Civil War
n. French Protestants
o. Plan to unify Spanish administration
p. Representative body minus royalists and Presbyterians
q. Series of conflicts over central governments' ambitions
r. Fee making purchased offices hereditary

CHRONOLOGICAL DIAGRAM

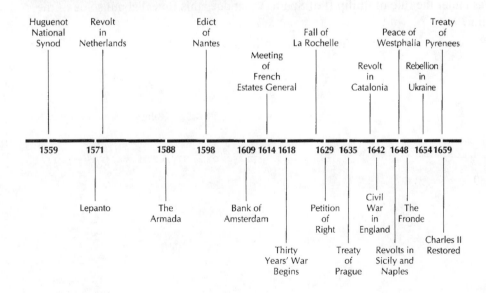

MAP EXERCISES

1. Indicate those areas under the rule of Philip II of Spain. What does this reveal about some of the problems facing him?
2. Indicate areas of Protestant resistance to the religious policies of Philip II.

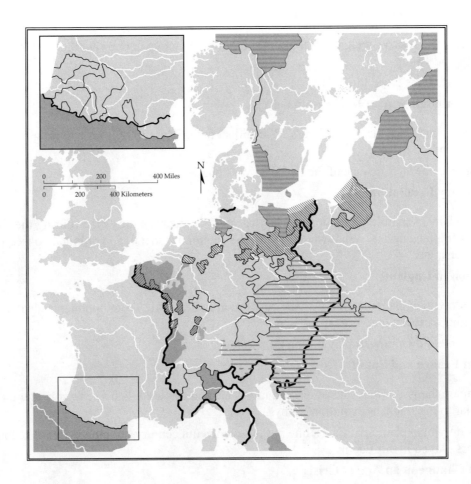

1. Indicate the lands controlled by the Austrian Habsburgs, the Spanish Habsburgs, France, Brandenburg-Prussia, and Sweden in 1660.

PROBLEMS FOR ANALYSIS

I. Rivalry and War in the Age of Philip II

1. Evaluate the relative weight of religious and political factors underlying the wars between the 1560s and 1640s.
2. What problems faced Philip II?

II. From Unbounded War to International Crises

1. In what ways does the Thirty Years' War reflect the mixture of politics and religion in war during this period?
2. What was the significance of the Peace of Westphalia?

III. The Military Revolution

1. What changes in military equipment, tactics, and organization took place during the sixteenth and seventeenth centuries? What were some of the economic, social, and political consequences of these changes?

IV. Revolution in England

1. How do you explain the civil war and revolution in England? Were opponents clearly split along religious or social lines?
2. What role did Cromwell play during this period?

V. Revolts in France and Spain

1. Compare the issues that caused discord in France and England. Is it fair to consider the Fronde similar to the English revolution? Why?
2. How do you explain Spain's decline in the seventeenth century? Evaluate the role economic factors played in this decline.

VI. Political Change in an Age of Crisis

1. What was so unusual about the United Provinces? How would you explain its rapid economic and cultural success?
2. How was Sweden able to rise from a second-rate power to a position of dominance in the Baltic during the seventeenth century? What policies did Gustavus Adolphus follow to this end?
3. Compare the general nature of constitutional settlements reached throughout Europe at mid-century. In what areas did monarchical power prevail? In what areas was monarchical power weakened?

SPECULATIONS

1. How might Charles I have prevented the English Civil War and revolution?
2. As a political leader in the early seventeenth century, would you use religion for your own ends? What are the dangers in doing or not doing this?
3. As adviser to Philip II of Spain, what policies would you recommend to prevent the eventual decline of Spanish power? Why?

TRANSITIONS

In "Economic Expansion and a New Politics," the economic advances during the late-fifteenth century and the sixteenth century, the process of European expansion, and the growth of political authority around the "new monarchs" of Western Europe were examined.

In "War and Crisis," we see the period between the 1560s and 1650s dominated by war and internal revolt. Politics and religion entwined in these upheavals, above all in the civil wars in England and France and the Thirty Years' War in the Holy Roman Empire. The period ends with a relative sense of constitutional and military stability in most places, which was to last for some time.

In "Culture and Society in the Age of the Scientific Revolution," the social, cultural, and intellectual patterns paralleling these political and international trends during the sixteenth and seventeenth centuries will be explored.

ANSWERS

Self Test

1b; 2c; 3d; 4a; 5d; 6b; 7a; 8d; 9c; 10c; 11b; 12c; 13b; 14d; 15d; 16c; 17b; 18d; 19a; 20a; 21c; 22d

Guide to Documents

I-1a; II-1b; III-1d

Significant Individuals

1r; 2j; 3m; 4f; 5o; 6i; 7q; 8p; 9a; 10e; 11d; 12n; 13h; 14l; 15c; 16g; 17b; 18k

Identification

1b; 2n; 3i; 4g; 5l; 6a; 7d; 8m; 9e; 10f; 11p; 12j; 13r; 14h; 15k; 16c; 17o; 18q

SIXTEEN
CULTURE AND SOCIETY IN THE AGE OF THE SCIENTIFIC REVOLUTION

MAIN THEMES

1. Major breakthroughs in physics, astronomy, mathematics, and anatomy resting on the new scientific principles of reason, doubt, observation, generalization, and testing by experiment overturned accepted ideas about nature and laid the foundations for modern science.

2. Cultural styles evolved, from the distortion of Mannerism to the drama of the Baroque and the discipline of Classicism.

3. Seventeenth-century society was hierarchical, although mobility was becoming increasingly common in the higher orders.

4. The traditional village was changing and being pulled into the activities of the territorial state, a process that often involved wrenching dislocations for the people involved.

5. While general attitudes were marked by belief in magic and mystical forces, these came to have less impact on public policy and the general culture as educated opinion became increasingly skeptical of them.

6. Both elite and popular culture generally followed a trajectory from passion and turmoil to restraint and order similar to the transition in politics and international relations during the same period.

OUTLINE AND SUMMARY

I. Scientific Advance from Copernicus to Newton

1. Origins of the Scientific Revolution
2. The Breakthroughs
3. Kepler and Galileo Address the Uncertainties
4. The Climax of the Scientific Revolution: Isaac Newton

II. The Effects of the Discoveries

1. A New Epistemology
2. The Wider Influence of Scientific Thought
3. Bacon and Descartes
4. Pascal's Protest Against the New Science
5. Science Institutionalized

III. Literature and the Arts

1. Unsettling Art
2. Unsettling Writers
3. The Return of Assurance in the Arts
4. Stability and Restraint in the Arts

IV. Social Patterns and Popular Culture

1. Population Trends
2. Social Status
3. Mobility and Crime
4. Change in Villages and Cities
5. Belief in Magic and Rituals
6. Forces of Restraint

SELF TEST

1. All of the following helped set the scientific revolution in motion EXCEPT
   a. inaccuracies in and inconsistencies among ancient authorities.
   b. magical beliefs that emphasized simple, comprehensive keys to nature.
   c. belief in the importance of observation and development of instruments.
   d. changes in Christianity that focused on its metaphorical rather than literal truth.

2. Which of the following was NOT one of the important scientific breakthroughs in the sixteenth century?
   a. Vesalius' anatomical studies.
   b. Copernicus' astronomy.
   c. Kepler's laws of planetary motion.
   d. Tycho Brahe's observations of the heavens.

3. Which of the following were NOT critical developments in astronomy and physics?
   a. Galileo's concept of inertia and his observations of the moons of Jupiter.
   b. Kepler's laws of planetary motion and his synthesis of them with terrestrial physics.
   c. Descartes development of analytic geometry and distinction between weight and mass.
   d. Newton's development of calculus and his three laws of motion.

4. Newton's work was the culmination of the scientific revolution because
   a. it resolved the outstanding problems in both physics and astronomy.
   b. it refuted Descartes' theoretical approach to scientific knowledge.
   c. it reconciled science with Christianity as it was then understood.
   d. it made further work unnecessary for the next 150 years.

5. The new epistemology of science involved all of the following EXCEPT
   a. reliance on experience and reason rather than authority.
   b. testing of an hypothesis by observation, generalization, and experimentation.
   c. rejection of Occam's theory that the simplest explanation is best.
   d. use of numerical data to develop mathematical laws.

6. Wider acceptance of scientific thought came when the educated public
    a. became convinced that science offered certainty.
    b. accepted that science cannot promise certainty.
    c. made the effort to follow the intricacies of scientific debate.
    d. became enamored of the charismatic figures of science.

7. Bacon and Descartes complemented each other because
    a. they had long been ardent admirers of each other.
    b. Bacon emphasized experiment and inductive reasoning while Descartes emphasized deductive analysis.
    c. Bacon was able to influence his fellow Englishmen, while Descartes worked on the French.
    d. they both were able to refute Newton's work by approaching it from opposite points of view.

8. Blaise Pascal is important because
    a. he promoted a reconciliation of Catholic faith with the new science.
    b. he kept European civilization from accepting the mechanical world view.
    c. he was the first accomplished scientist to focus on the limitations of science.
    d. he was able to undercut some of the extreme claims of science on a scientific basis.

9. All of the following were examples of popular enthusiasm for science in the late seventeenth century EXCEPT
    a. royal patronage of scientific societies.
    b. the use of science as an aristocratic amusement.
    c. popular attendance at autopsies.
    d. the widespread practice of *charivaris*.

10. The primary impulse behind Mannerism was
    a. distortion.
    b. escapism.
    c. mysticism.
    d. singularity.

11. Michel de Montagne created the literary form known as the
    a. essay.
    b. novel.
    c. reflections.
    d. epigram.

12. Cervantes and Shakespeare had in common that they both
    a. were Englishmen.
    b. had essentially optimistic outlooks.
    c. reflected the stresses of their times.
    d. rejected the hierarchy of society.

13. The Baroque style was found particularly in Catholic countries because
    a. it supported the Counter Reformation.
    b. only they had the wealth to support it.
    c. Protestants preferred more flamboyant styles.
    d. only there were artists with necessary skills found.

14. The Classical style was differentiated from the Baroque because
    a. it was characterized by restraint and discipline.
    b. it aimed at grandiose effects.
    c. it was found mainly in Protestant countries.
    d. it included art forms beyond painting.

15. The number of Europeans rose only slightly in the seventeenth century for all the following reasons EXCEPT
    a. economic pressures caused people to marry late, which reduced the number of babies they could have.
    b. the Thirty Years' War killed millions of Germans and disrupted the European economy.
    c. plagues drove the number of Spaniards down from 10 million in 1600 to 6 million in 1700.
    d. the English and Dutch populations only recovered after 1680, accounting for the little increase there was.

16. Seventeenth century society was characterized by
    a. impenetrable class barriers.
    b. relative egalitarianism.
    c. significant mobility.
    d. decreasing stratification.

17. Life was generally becoming harder for the peasantry for all of the following reasons EXCEPT
    a. taxes were rising.
    b. rents and other dues were increasing.
    c. food prices were stabilizing.
    d. there was no escape from the farm.

18. All of the following changes were taking place in traditional villages EXCEPT
    a. differences in wealth among the peasants were increasing.
    b. government officials were eroding village self-government.
    c. government welfare programs were enticing peasants off the land.
    d. noble landlords were ceasing to pay attention to their villagers' lives.

19. In contrast to the villages, life in the cities was
    a. secure.
    b. impersonal.
    c. dull.
    d. fragile.

20. All of the following first became common in cities in the late seventeenth century EXCEPT
    a. weekly newspapers.
    b. coffeehouses.
    c. actresses.
    d. books.

21. Urban and rural culture were distinguished by all of the following EXCEPT
    a. literacy rates.
    b. magical beliefs.
    c. types of recreation.
    d. visibility of religiosity.

22. The witch craze was caused by all of the following EXCEPT
    a. an intention to persecute innocent people.
    b. popular fears that bad effects can be willed.
    c. an official desires to root out agents of evil.
    d. a fear of women who seemed too potent.

23. The decline of witch-hunting occurred for all of the following reasons EXCEPT
    a. official recognition that it was disruptive and dangerous.
    b. the rising cultural weight of cities, where events are more subject to rational human control.
    c. religious changes that de-emphasized magic and disapproved of passionate popular activities.
    d. the rapid spread of the scientific world view to all classes of society.

GUIDE TO DOCUMENTS

I. Galileo and Kepler on Copernicus

1. Aside from their personalities, what seems in these passages to explain why Kepler is more open than Galileo?
    a. He is more convinced Copernicus is right.
    b. His is the more powerful mind.
    c. He lives in a more tolerant country.
    d. He thinks fewer scholars still oppose Copernicus.

2. What does this reveal about the potential strength of the new scientific community in Europe?

II. A Witness Analyzes the Witch Craze

1. According to Linden, what was the primary dynamic driving the witch-hunts?
   a. Witches caused bad harvests; so the people and government prosecuted them.
   b. The irrational fears of the populace drove them to carry out mob justice.
   c. Corrupt officials created a scandal about witches in order to make a profit.
   d. Corrupt officials exploited popular fears in order to make a profit.

2. What were some of the forces that led to the decline of the witch-hunts?

IDENTIFICATION

1. alchemy (AL-ke-mē)
2. scientific method
3. three laws of motion
4. mechanism
5. skepticism
6. Royal Society of London
7. *Philosophical Transactions*
8. *Principia*
9. Baroque (ba-RŌK)
10. seigniorial reaction (sēn-YOR-ē-al)
11. witch craze

a. Core principals of Newton's physics
b. First scientific journal
c. Movement by landlords to squeeze more out of peasants
d. Campaign to eradicate evil magicians
e. Short name of Newton's primary work
f. Philosophy of doubt
g. Cultural style emphasizing grandeur and excitement
h. Rituals believed to transform one substance into another
i. Theory that universe is a machine subject to physical laws
j. Process of hypothesis, observation, generalization, and experimentation
k. Group that promoted scientific research

CHRONOLOGICAL DIAGRAM

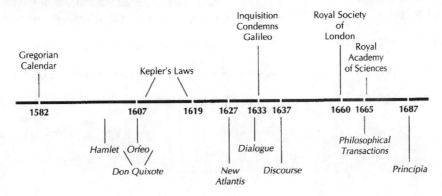

SIGNIFICANT INDIVIDUALS

1. Andreas Vesalius (vi-SĀ-lē-us)
2. Nicolaus Copernicus (kō-PUR-ni-Kus)
3. William Harvey
4. Johannes Kepler (KEP-luhr)
5. Galileo Galilei (gal-uh-LĒ-ō gal-uh-LĀ)
6. Isaac Newton
7. Francis Bacon
8. René Descartes (dā-KART)
9. Blaise Pascal (pas-KAL)
10. El Greco (GREK-ō)
11. Michel de Montaigne (mon-TĀN)
12. Miguel de Cervantes (ser-VAN-tēz)
13. William Shakespeare
14. Caravaggio (ka-ra-VAD-jō)
15. Peter Paul Rubens
16. GianLorenzo Bernini (JĒ-an LO-ren-zō ber-NĒ-nē)
17. Rembrandt van Rijn (ryn)
18. Caludio Monteverdi (mon-tā-VER-dē)
19. Nicholas Poussin (poo-san)
20. Pierre Corneille (kor-NĀ
21. Jean Racine (ra-SĒN)

a. Scientist who formulated the laws of motion (1642-1727)
b. Astronomer who determined that the planets' orbits are elliptical (1571-1630)
c. Philosopher who emphasized deduction (1596-1650)
d. French Classical painter (1594-1665)
e. Originator of Baroque style (1571-1610)
f. Premier English playwright (1564-1616)
g. Baroque sculptor and architect (1598-1680)
h. Philosopher who emphasized induction (1561-1626)
i. Dutch painter who transcended style (1606-1669)
j. Leading Flemish Baroque painter (1577-1640)
k. Anatomist who discovered that the heart pumps blood (1578-1626)
l. French playwright uneasy with Classicism who wrote *Le Cid* (1606-1684)
m. Creator of opera and orchestra (1567-1643)
n. Leading skeptic and essayist (1533-1592)
o. Model French Classical playwright (1639-1699)
p. Astronomer who said the Earth circles the Sun (1473-1543)
q. Author of *Don Quixote* (1547-1616)
r. Scientist who created the concept of inertia, built a telescope, and discovered Jupiter's moons (1564-1643)
s. Anatomist who found Galen inaccurate (1514-1564)
t. Scientist who questioned science (1623-1662)
u. Leading Mannerist painter (1548?-1625?)

MAP EXERCISE

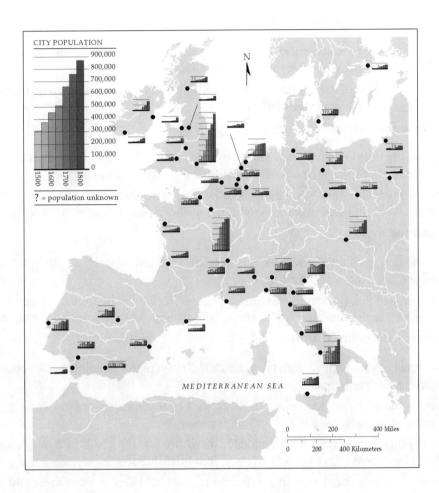

1. Label the five cities that grew most rapidly between 1500 and 1800.
2. Indicate and label Europe's eight largest cities in 1700.

PROBLEMS FOR ANALYSIS

I. Scientific Advance from Copernicus to Newton

1. Explain the origins of the scientific revolution. Were the theories of the ancient Greeks a hindrance or a support? What role did magical beliefs play?
2. What was the essence of the conflict between Galileo and the Church? Do you think it was in the Church's interest to condemn Galileo? Why?

II. The Effects of the Discoveries

1. Use examples to demonstrate the principles of the scientific method. How does the scientific method differ from earlier methods of obtaining and verifying knowledge?
2. Compare the methods emphasized by Francis Bacon, René Descartes, and Isaac Newton. Do you think that Blaise Pascal would disagree with the methods and concerns of these men? Why?

III. Literature and the Arts

1. Compare the Baroque and Classical styles. In what ways did they reflect other developments in sixteenth- and seventeenth-century Europe?
2. It has been argued that Cervantes and Shakespeare reflected the historical concerns of their own societies as well as timeless human concerns. Use examples to support this argument.

IV. Social Patterns and Popular Culture

1. How was seventeenth-century society organized? Compare the possibilities for social mobility among various social groups.
2. What were some of the demographic characteristics of seventeenth-century society? How do you explain population patterns in the seventeenth century?
3. How did popular culture reflect a dependence on nature and the conditions of life among the peasantry? How does the great witch craze fit into this situation?
4. What were the causes for change in the traditional village? Were most of these causes internal, or were they a result of intrusions from the outside?

SPECULATIONS

1. The scientific revolution profoundly changed the ways in which people thought. It was difficult for many to accept this change. Today scientific ways of thinking are as accepted and taken for granted as traditional ways of thinking in the sixteenth century. What might a future change in the ways of thinking be like, and do you think such ways of thinking would be accepted without too much difficulty?
2. Imagine what a debate between Galileo and the head of the Inquisition would be like? Write a script for this exchange as if you were writing a play, including both their words and their expressions and gestures.
3. Are there any parallels between the great witch craze of the seventeenth century and more recent historical occurrences? Explain.

TRANSITIONS

In "War and Crisis," a period of violence and upheaval marked by unusually brutal warfare was examined. It was not until the mid-seventeenth century that the violence subsided and a new sense of order was attained.

In "Culture and Society in the Age of the Scientific Revolution," cultural and social patterns are shown to reflect this progression from uncertainty to stable resolution. This is clearest in the triumph of the scientific revolution—the revolutionary discoveries of a handful of men who laid the foundations for modern science—but it is also apparent in the evolution from Mannerism to the Classical style and in the increased control over people's lives gained by central governments. The upper classes throughout Europe benefited most from these trends.

In "The Emergence of the European State System," the course of political history during the second half of the seventeenth century will be explored. In this period absolutist kings continue the process of state building within a European society dominated by the aristocracy.

ANSWERS

Self Test

1d; 2c; 3b; 4a; 5c; 6a; 7b; 8c; 9d; 10b; 11a; 12c; 13a; 14a; 15d; 16c; 17d; 18c; 19b; 20d; 21b; 22a; 23d

Guide to Documents

I-1c; II-1d

Significant Individuals

1s; 2p; 3k; 4b; 5r; 6a; 7h; 8c; 9t; 10u; 11n; 12q; 13f; 14e; 15j; 16g; 17i; 18m; 19d; 20l; 21o

Identification

1h; 2j; 3a; 4i; 5f; 6k; 7b; 8e; 9g; 10c; 11d

SEVENTEEN
THE EMERGENCE OF THE EUROPEAN STATE SYSTEM

CHAPTER HIGHLIGHTS

1. Louis XIV—by making his court at Versailles the center of society and by building the state's power through financial, domestic, and military policies—epitomized the absolutist monarchs of the late seventeenth century.
2. In related ways, absolutism grew in Austria, Prussia, Russia, and, to a lesser extent, Spain.
3. The governments of England, the United Provinces, Sweden, and Poland were dominated by aristocrats or merchants. With the exception of England, these countries suffered a decline in power and influence.
4. During the eighteenth century the states of Europe competed for power within a system that created a balance of power.

CHAPTER OUTLINE

I. The Creation of Absolutism in France
1. The Rule of Louis XIV
2. Government
3. Foreign Policy
4. Domestic Policy
5. The End of an Era
6. France after Louis XIV

II. Other Patterns of Absolutism
1. The Habsburgs at Vienna
2. The Hohenzollerns at Berlin
3. Rivalry and State Building
4. The Prussia of Frederick William I
5. Frederick the Great
6. The Habsburg Empire
7. The Habsburgs and Bourbons at Madrid
8. Peter the Great at St. Petersburg

IV. Alternatives to Absolutism

1. Aristocracy in the United Provinces, Sweden, and Poland
2. The Triumph of the Gentry in England
3. Politics and Prosperity
4. The Growth of Stability
5. Contrasts in Political Thought

V. The International System

1. Diplomacy and Warfare
2. Armies and Navies
3. The Seven Years' War

SELF TEST

1. Louis XIV's court at Versailles was designed to serve all of the following purposes EXCEPT
    a. to impress people with his wealth, power, and refinement.
    b. to insulate the court from the turmoil of the capital city.
    c. to serve as a final defensive bastion in case of invasion.
    d. to detach nobles from their traditional bases of power in the provinces.

2. Developing the country's bureaucracy gave Louis increased ability to do all of the following EXCEPT
    a. expand and control the armed forces.
    b. formulate and execute laws.
    c. collect and disburse revenue.
    d. disenfranchise the traditional nobility.

3. Louis' foreign policy resulted in
    a. decisive victories.
    b. greater gains than losses.
    c. only marginal gains at great cost.
    d. great losses of territory and resources.

4. Louis XIV's domestic policy included all of the following EXCEPT
    a. fostering manufacturing, agriculture, and trade.
    b. expelling the Huguenots and suppressing Jansenism.
    c. quashing legal protests and crushing peasant rebellions.
    d. transforming the aristocracy into a compulsory service class.

5. From late in Louis XIV's reign through the middle of Louis XV's, conditions in France generally went
    a. from bad to worse.
    b. from good to bad.
    c. from bad to better.
    d. from good to better.

6. After Louis XIV, the French monarchy was troubled by all of the following EXCEPT
    a. renewed competition from aristocrats (especially in the parlements).
    b. financial instability (thanks to exemptions from taxes enjoyed by the privileged).
    c. incessant warfare (thanks to Louis XV's dynastic ambitions in Spain).
    d. political weakness (except during the ministry of Cardinal Fleury).

7. Leopold I of Austria's rule was characterized by all of the following EXCEPT
    a. establishment of a Versailles-like palace at Schönbrunn.
    b. reliance on aristocrats to help rule nationally and locally.
    c. strong efforts to make Imperial rule effective in Germany.
    d. significant expansion to the southeast at Ottoman expense.

8. Frederick William made Brandenburg-Prussia into a power in Germany by all of the following EXCEPT
    a. building a strong army, which rose from 8,000 men in 1648 to 22,000 in the 1650s (and 200,000 in 1786).
    b. allying with the nobles, who got control of the peasants and through serfdom made their estates profitable.
    c. organizing the state to sustain the army by having officers run the treasury and local administration.
    d. gaining the title of King in Prussia and making Berlin into a cosmopolitan social and cultural center.

9. International competition spurred internal state building because
    a. an efficient bureaucracy, prosperous economy, and stable society were the foundations of military power.
    b. as conquered peoples came under different rulers, they made use of the best aspects of each government.
    c. larger powers were able to swallow up smaller states wholesale, and had to digest and integrate them.
    d. rulers vied for the distinction of ruling the most fortunate state by best serving the needs of their people.

10. Frederick William I did all of the following EXCEPT
    a. increase the size of the army.
    b. improve the quality of the officers.
    c. wear an army uniform at all times.
    d. fight a war.

11. Frederick the Great was all of the following EXCEPT
    a. an outstanding general.
    b. a God-fearing German Protestant.
    c. a composer, poet, and philosopher.
    d. a ruthless statesman.

12. The Habsburgs faced all of the following difficulties in forging their empire EXCEPT
    a. it was made up of socially and culturally diverse territories united only by the dynasty that ruled them.
    b. the local nobles in the different territories jealously defended, and tried to extend, their traditional rights.
    c. Prussia, France, Spain, and Bavaria tried to take advantage of the succession of Maria Theresa, a woman.
    d. they lost a number of provinces because Hungarian troops and British gold proved insufficient support.

13. Maria Theresa accomplished all of the following EXCEPT
    a. expanding Austria's tax base.
    b. founding new monasteries.
    c. reforming the administration.
    d. modernizing the army.

14. Spain remained an important international player in the eighteenth century because of its
    a. large population.
    b. powerful navy.
    c. victorious army.
    d. economic strength.

15. Peter the Great accomplished all of the following during his reign EXCEPT
    a. establishing Russia as a major presence in the Black Sea.
    b. beginning the westernization of Russia's economy and society.
    c. taking control of the Church and ignoring representative institutions.
    d. reducing the peasants to the level of serfs and forcing the nobles to serve the state.

16. Holland, Sweden, and Poland in the eighteenth century had in common that
    a. they lost out because they failed to modernize their political systems.
    b. they lost out to neighbors who mobilized superior national power.
    c. they lost out because they overextended themselves when successful.
    d. they lost out when they could not successfully defend their territory.

17. The Glorious Revolution confirmed the gentry's control of England in all of the following ways EXCEPT
    a. it reconfirmed that the monarchy did not have the power to defy Parliament.
    b. it established that the king's ministers were also responsible to Parliament.
    c. William III accepted legal restraints on his power in exchange for the crown.
    d. James II took refuge with Louis XIV, identifying absolutism with the enemy.

18. All of the following both contributed to and resulted from England's economic prosperity EXCEPT
    a. the success of the Bank of England.
    b. the rise of the Navy.
    c. Tory dominance in politics.
    d. overseas expansion.

19. All of the following changes took place in eighteenth century Britain EXCEPT
    a. The House of Commons came to be dominated by landowners and leading townsmen.
    b. Britain created a bureaucratized state with a standing army and expanding navy.
    c. Executive power came to be directed by a cabinet of ministers responsible to Parliament.
    d. Dominance in setting foreign policy shifted from the landholders to the commercial elite.

20. The main difference between Hobbes and Locke was that Locke argued
    a. people in nature have liberty but not security.
    b. government is created by a contract to secure people's lives and property.
    c. the sovereign is a party to the contract, and may be overthrown if he breaks it.
    d. if the sovereign is overthrown, people revert to a state of nature.

21. During the eighteenth century, international relations came to be dominated by all of the following EXCEPT
    a. the impersonal interests of the states rather than the dynastic concerns of the rulers.
    b. an aristocratic, cosmopolitan, French-speaking corps of professional diplomats.
    c. the belief that any means were justified in the pursuit of power.
    d. a balance of power that protected every state's security.

22. During the eighteenth century, all of the following were true of armies and navies EXCEPT
    a. they became much larger and more expensive.
    b. officers and men became more professional.
    c. fighting was restrained by the costs of equipment and trained manpower.
    d. weapons and tactics changed radically as new technologies were developed.

23. The Seven Years' War was fought by
    a. England and Prussia against Austria, France, and Russia.
    b. England and Russia against Austria, France, and Prussia.
    c. France and Austria against England, Prussia, and Russia.
    d. France and Russia against England, Austria, and Prussia.

GUIDE TO DOCUMENTS

I. Louis XIV on Kingship

1. What, for Louis XIV, is the ultimate source of monarchical authority?
    a. A covenant with the people.
    b. Divine right.
    c. The realities of power.
    d. The king's virtue.

2. According to Louis XIV, in what ways should the monarch act?

II. Locke on the Origins of Government

1. According to Locke, why do men exit the state of nature and form a society with a government?
   a. To maintain their liberty.
   b. To take control of others.
   c. To safeguard their property.
   d. To achieve the common good.

2. What, according to Locke, are the limits of society's power over the individual?

III. Maria Theresa in Vehement Mood

1. How does Maria Theresa justify her diplomatic realignment?
   a. She saw her opportunity and took it.
   b. The good of the Austrian people.
   c. England betrayed her first.
   d. Alliance with Russia was more important than with England.

2. What, according to Maria Theresa, are Austria's proper reasons of state?

IDENTIFICATION

1. Versailles (ver-SĪ)
2. Grand Alliance
3. Peace of Utrecht (YOO-trekt)
4. Act of Toleration
5. *vingtième* (van-tā-em)
6. Junkers (YOONG-kers)
7. the War Chest
8. Schönbrun (SCHOEN-brun)
9. Pragmatic Sanction
10. Bill of Rights
11. Whigs (hwigs)
12. Tories
13. *tabula rasa* (TAB-yoo-la RA-sa)
14. reasons of state
15. the diplomatic revolution

a. Prussian nobles
b. Diplomatic agreement recognizing Maria Theresa
c. English party for royal power and against war
d. Needs of the government that override other concerns
e. Leopold I's palace
f. Treaty that ended the War of Spanish Succession
g. English party against royal power and for war
h. Theory that a newborn baby's brain is a blank slate
i. Law establishing religious freedom in England
j. Louis XIV's palace
k. New set of alliances and enmities
l. Louis XIV's opponents in War of Spanish Succession
m. Law on succession, Parliament's powers, and civil rights
n. French tax intended to tap the wealth of all classes
o. Prussian treasury department

CHRONOLOGICAL DIAGRAM

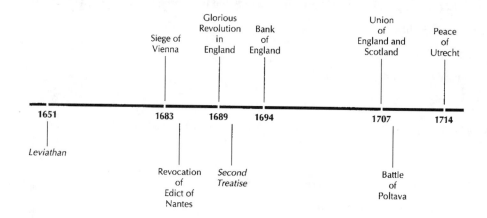

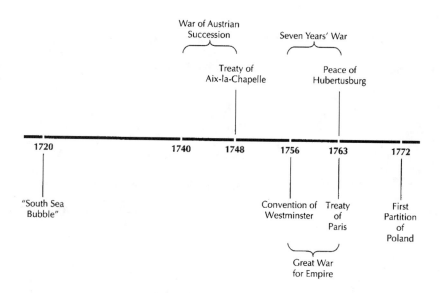

SIGNIFICANT INDIVIDUALS

1. Louis XIV
2. Jean-Baptiste Colbert (kol-BAIR)
3. Marquis de Louvois (lou-VWAH)
4. Louis XV
5. Cardinal Fleury (flue-RĒ)
6. Frederick William, the Great Elector
7. Frederick William I
8. Frederick the Great
9. Maria Theresa
10. Peter the Great
11. Charles XII
12. Charles II
13. James II
14. William III
15. George I
16. Robert Walpole
17. William Pitt
18. Thomas Hobbes (hobz)
19. John Locke (lok)

a. King of France known as the Sun King (r.1643-1715)
b. Empress of Austria who lost Silesia (r. 1740-1780)
c. Prussian king who built up army but didn't fight (r.1701-1740)
d. Louis XIV's advisor to who focused on war (1641-1691)
e. Louis XIV's advisor who focused on money (1619-1683)
f. Leader who made Brandenburg-Prussia a power (r.1640-1688)
g. Pessimistic political philosopher (1588-1679)
h. Restored king of England (1660-1685)
i. "First" prime minister of England (r.1721-1742)
j. Chief advisor to Louis XV (r.1720-1740)
k. Dutch king of England (r.1689-1702)
l. Philosopher of "Life, Liberty, and Property" (1632-1704)
m. Parliamentary leader in Seven Years' War (r.1708-1778)
n. Tsar who modernized Russia (1682-1725)
o. First Hanoverian king of England (r.1714-1727)
p. Aggressive and lucky Prussian king (r.1740-1786)
q. Swedish king who lost empire (1697-1718)
r. French monarch in eighteenth century (1715-1774)
s. English king until Glorious Revolution (1685-1688)

MAP EXERCISE

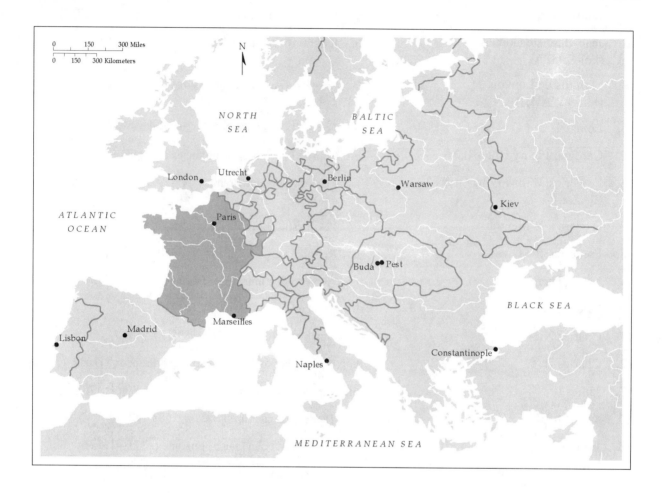

1. Label the main countries and empires in Europe in 1715.
2. Indicate areas where Louis XIV made efforts to expand France.

PROBLEMS FOR ANALYSIS

I. The Creation of Absolutism in France

1. In what ways did the policies of Louis XIV build the state's power? What evidences success or failure of these policies?
2. Which view of Louis XIV's reign do you find more persuasive, Albert Sorel's or John Rule's? Why?

II. Austrian and Prussian Absolutism

1. Explain the relative success of Prussia during the eighteenth century.
2. How did Austria centralize state power during this period?

III. Absolutism in Spain and Russia

1. Compare Russian absolutism to that of France and Prussia. In what ways was it similar to each? How was it different from both?

IV. Alternatives to Absolutism

1. What developments support the argument that during the late seventeenth century the gentry triumphed in England?
2. Compare the decentralization of government and society that occurred in the United Provinces, Sweden, and Poland during the late seventeenth and early eighteenth centuries.

V. The International System

1. Compare the nature of war and diplomacy during the eighteenth century with that of the late sixteenth and early seventeenth centuries.
2. What role did "reasons of state" and dynastic interests play in the wars, diplomacy, and internal policies of European states during the eighteenth century? Give examples.

SPECULATIONS

1. First as an aristocrat, and second as a merchant, what were the advantages and disadvantages of living in a country dominated by an absolutist monarch?
2. How did Hobbes and Locke disagree with each other?
3. How might Machiavelli view political, diplomatic, and military developments during the eighteenth century? How might Hobbes?

TRANSITIONS

In "Culture and Society in the Age of the Scientific Revolution," the fundamental scientific discoveries and the cultural creations of this period were examined. These achievements contributed to a sense of order by the mid-seventeenth century.

In "The Emergence of the European State System," the quest for order remained the underlying concern throughout Europe. Absolutist kings, epitomized by Louis XIV, rose to prominence. With the exception of England, those states that failed to focus power on the monarch declined. The aristocracy, to varying degrees, dominated the new, powerful governmental administration as needed allies and agents of absolute monarchs or as direct controllers of events. During the eighteenth century international competition was reflected in efforts to further build the state internally.

In "The Wealth of Nations," the new social and economic developments as well as the development of eighteenth-century empires will be examined.

ANSWERS

Self Test

1c; 2d; 3c; 4d; 5c; 6c; 7c; 8d; 9a; 10d; 11b; 12d; 13b; 14b; 15a; 16b; 17b; 18c; 19a; 20c; 21d; 22d; 23a

Guide to Documents

I-1b; II-1c; III-1c

Significant Individuals

1a; 2e; 3d; 4r; 5j; 6f; 7c; 8p; 9b; 10n; 11q; 12h 13s; 14k; 15o; 16i; 17m; 18g; 19l

Identification

1j; 2l; 3f; 4i; 5n; 6a; 7o; 8e; 9b; 10m; 11g; 12c; 13h; 14d; 15k

EIGHTEEN
THE WEALTH OF NATIONS

CHAPTER HIGHLIGHTS

1. The eighteenth century, particularly after 1730, was a period of population growth, gently rising prices, and protoindustrialization.
2. Building upon conditions favorable to industrialization, England initiated a new economic order. New tools, machines, and sources of energy enormously increased the productivity of labor, particularly in cotton textiles.
3. Compared to England, the legal and social conditions in France and other continental countries were less conducive to economic development, and they experienced real but less dramatic growth.
4. New agricultural techniques and enclosures in Britain, and to a much lesser extent in certain areas of the continent, enabled the countryside to supply industrial towns with food and labor as well as capital and growing markets.
5. The conditions of peasants and serfs in Eastern Europe were worse than in Western Europe.
6. France and Britain became the dynamic maritime powers, competing and warring for lucrative colonies and trade.
7. The primary sources of their wealth and power were the slave plantations in the Caribbean, which were sustained by the "triangular trade" across the Atlantic.

CHAPTER OUTLINE

I. Demographic and Economic Growth
1. A New Demographic Era
2. Profit Inflation: The Movement of Prices
3. Protoindustrialization

II. The New Shape of Industry
1. Toward a New Economic Order
2. The Roots of Economic Transformation in England
3. Cotton: The Beginning of Industrialization

III. Innovation and Tradition in Agriculture
1. Convertible Husbandry
2. The Enclosure Movement in Britain
3. Serfs and Peasants on the Continent

IV. Eighteenth-Century Empires

1. Mercantile and Naval Competition
2. The Profits of Empire
3. Slavery, the Foundation of Empire
4. Mounting Colonial Conflicts
5. The Great War for Empire
6. The British Foothold in India

SELF TEST

1. During the eighteenth century, Europe's population grew by
    a. 25%
    b. 50%
    c. 75%
    d. 100%

2. Europe's population rose primarily because of
    a. declining death rates due to better food supply.
    b. declining death rates due to improved medicine and hygiene.
    c. increasing birth rates due to increased economic opportunities.
    d. increasing birth rates due to a mild improvement in average climate.

3. How did the population rise affect the scope of economic activity during the eighteenth century?
    a. Increasing competition for resources depressed the economy.
    b. Increasing demand for good and services stimulated the economy.
    c. Increasing competition and increasing demand canceled each other, so there was little net effect.
    d. Europe's surplus population emigrated to the colonies, making up the bulk of the labor force in them.

4. Europe's economic growth had which of the following effects?
    a. Rents rose while wages fell, benefiting landlords and businessmen and hurting farmers and laborers.
    b. Rents rose while wages fell, benefiting farmers and laborers and hurting landlords and businessmen.
    c. Rents fell while wages rose, benefiting landlords and businessmen and hurting farmers and laborers.
    d. Rents fell while wages rose, benefiting farmers and laborers and hurting landlords and businessmen.

5. Protoindustrialization had all of the following effects EXCEPT
    a. economically, it strengthened marketing networks, spurred capital accumulation, and stimulated demand.
    b. socially, it improved the peasants' lives while teaching them about money and industrial production.
    c. demographically, it loosened restraints on marriage and births, which led to immigration into the cities.
    d. technologically, it stimulated innovations that increased productivity that sustained continuous growth.

6. Use of better tools and the new sources of energy had all of the following effects EXCEPT

    a. increasing the productivity of labor.

    b. transforming manufacturing.

    c. making work easier and the workday shorter.

    d. creating a new social institution: the factory.

7. All of the following were impediments to economic innovation EXCEPT

    a. small markets due to transportation limitations.

    b. demand skewed to finely crafted luxury items.

    c. entrenched economic privileges such as guilds and monopolies.

    d. *laissez-faire* attitudes among government administrators.

8. England satisfied all of the following preconditions for innovation and economic growth EXCEPT

    a. an excellent base of raw materials and transportation lines.

    b. a docile working population held firmly in place by *seigniorialism*.

    c. a uniform system of tariffs, laws, and standards.

    d. a relatively wealthy population and pool of potential entrepreneurs.

9. The techniques of mass production were first developed for

    a. iron making.

    b. making cotton goods.

    c. ship building.

    d. distilling rum from colonial sugar.

10. Well-to-do landlords were readier to experiment with agriculture than peasants because

    a. they were more intelligent than the peasants.

    b. their lives did not depend on it.

    c. aristocrats tended to be forward-looking.

    d. peasants had an unthinking resistance to change.

11. Enclosure had all of the following effects EXCEPT

    a. increasing the productivity of the land.

    b. destroying cooperative farming villages.

    c. driving many peasants into destitution.

    d. causing a massive rural depopulation.

12. Eastern Europe was distinguished from Western Europe by

    a. the continued existence of "feudal" relationships.

    b. the dominant division between nobles and serfs.

    c. the vigor of its urban commercial classes.

    d. the absolutist nature of its governmental structures.

13. The British and French competed to control all of the following areas EXCEPT
    a. the West Indies.
    b. East Africa.
    c. North America.
    d. India.

14. The triangular trade routes most consistently involved
    a. Europe.
    b. Africa.
    c. the West Indies.
    d. North America.

15. The West Indies were ideal colonies for all of the following reasons EXCEPT
    a. the slave system worked well since they were islands.
    b. they produced tropical crops difficult to grow elsewhere.
    c. they depended on exports from Europe and America.
    d. their climate was favorable for European habitation.

16. The European slave system had all of the following effects EXCEPT
    a. removing over 9,000,000 productive people from Africa.
    b. condemning some captives to death and the rest to a life of forced labor.
    c. creating luxuries and generating huge profits for the Europeans.
    d. populating Latin America with a predominantly black population.

17. Britain won its rivalry for dominance in the colonial world with France because
    a. Britain could focus on taking control of the seas and colonies, while France also had to fight in Europe.
    b. Britain was able to stay out of the Seven Years' War, while France fought in it.
    c. Britain's ally Russia was able to beat France, Austria, and Prussia.
    d. the French economy was more dependent on overseas trade, and thus more vulnerable in a naval war.

18. The Treaty of Paris did all of the following EXCEPT
    a. eliminate French power in North America.
    b. reduce French holdings in the West Indies.
    c. establish British military supremacy in India.
    d. remove France as a naval power of consequence.

19. Britain established control of India by all of the following EXCEPT
    a. moving aggressively to establish an overwhelming presence.
    b. exploiting divisions among Indian potentates.
    c. creating a class of landholders beholden to Britain.
    d. indoctrinating select Indians in British culture.

GUIDE TO DOCUMENTS

I. Laissez-Faire Ideology

1. According to Adam Smith, how does the pursuit of profits by the individual lead to promotion of the public good?

    a. The individual is best positioned to gauge the way in which his actions will affect the public good.

    b. The individual can best gauge what investment of resources will lead to the greatest return of value.

    c. An invisible hand informs him of the relative good that his actions will cause.

    d. An invisible hand guides him to forsake his own good in pursuit of the higher, public good.

2. What are the weaknesses of Smith's analysis?

II. Richard Arkwright's Achievement

1. According to the author Andrew Ure, the key problem in establishing the factory system was

    a. discerning how vastly productive human industry would become once mechanized.

    b. devising a self-acting mechanism for drawing out and twisting cotton into continuous thread.

    c. distributing different members of the mechanical apparatus into one co-operative body.

    d. inducing workers to labor at the pace dictated by the machines rather than their own comfort.

2. Ure describes Richard Arkwright in heroic terms. Do you think such a description is warranted. Why?

III. The Condition of the Serfs in Russia

1. What does this excerpt reveal about the relationship of nobles and serfs in Russia?

    a. All nobles treated their serfs like oxen.

    b. Nobles had arbitrary power over their serfs' lives.

    c. Serfs were incapable of productive work unless organized by the nobles.

    d. Modernization measures by progressive nobles would ultimately benefit the serfs.

2. What helps explain the power of the Russian nobility over the serfs?

IV. A British Defense of Slavery and the Plantation Economy

1. On what grounds does the author defend slavery?

    a. Biblical authority.

    b. "The white man's burden."

    c. Its value to England.

    d. The progress of man.

2. How does this defense of slavery and the plantation economy fit with the doctrines of mercantilism discussed in this chapter?

SIGNIFICANT INDIVIDUALS

1. Adam Smith
2. Charles Townshend
3. Jethro Tull
4. Richard Arkwright
5. Edmund Cartwright
6. James Watt
7. William Pitt

a. Early large scale agricultural innovator (1674-1741)
b. Inventor of the power loom (1780s)
c. Statesman who led Britain to rule the waves (1707-1778)
d. Agricultural innovator known as "Turnip" (1674-1738)
e. Developer of the steam engine (1780s)
f. Economist, author of *Wealth of Nations* (1723-1790)
g. Inventor of the water frame (1760s)

IDENTIFICATION

1. protoindustrialization
2. convertible husbandry
3. enclosure acts
4. peasant survival strategies
5. mercantilism
6. triangular trade
7. middle passage
8. The Great War for Empire
9. Treaty of Paris

a. The use of land for alternative crops rather than fallow
b. Laws ending open field system
c. Trade between Europe, Africa, and America
d. Doctrine of government intervention to foster trade
e. Slaves' voyage from Africa to America
f. Peace agreement that ended the Great War for Empire
g. The flowering of the "putting out" system
h. Colonial struggle between Britain and France
i. Control of sufficient land to meet basic economic needs

MAP EXERCISE

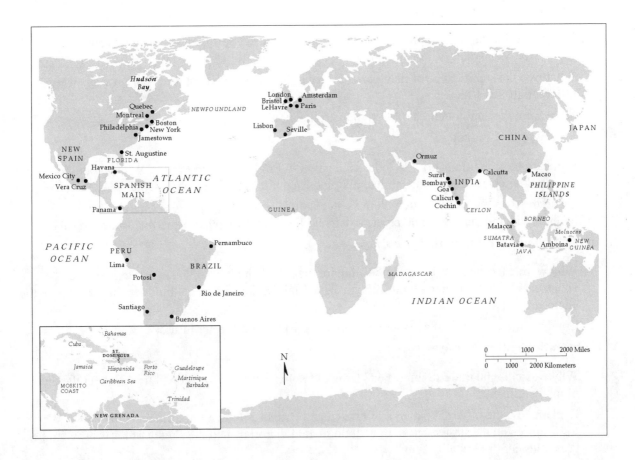

1. Indicate the English, French, Dutch, Spanish, and Portuguese possessions in the early eighteenth century.
2. Indicate the areas that changed hands in 1763.

PROBLEMS FOR ANALYSIS

I. Demographic and Economic Growth

1. How might the substantial population growth that occurred during the eighteenth century be explained? What factors do you think were most crucial? Why?
2. What was the significance of rising prices during the eighteenth century? In what ways were different groups affected?
3. What were the consequences of protoindustrialization?

II. The New Shape of Industry

1. Compare the conditions prior to industrialization in England and in France. Which conditions do you think are particularly crucial in explaining England's early industrial lead? Why?
2. Using the manufacture of cotton cloth as an example, explain the process whereby a new invention in an industry would create a demand for further innovations in that same industry.

III. Innovation and Tradition in Agriculture

1. Why were enclosures so important? Compare the social situation of areas that experienced enclosures with those that did not. What role did innovations in agricultural techniques play in the enclosure movement?
2. Compare the condition of serfs and peasants in Eastern and Western Europe.

IV. Eighteenth-Century Empires

1. Why was colonial trade so important? How did policies related to colonial trade reflect the theory of mercantilism?
2. Explain how the different patterns of French and British colonization influenced the Great War for Empire and the strategy used by Britain to gain victories. Did the Treaty of Paris prove to be a great blessing for Britain? Why?

SPECULATIONS

1. Describe the characteristics of a country ideally suited for rapid industrialization. What policies should the government of such a country follow to further facilitate rapid industrialization? Why?
2. Considering England and Europe during the eighteenth centuries, how might agricultural and entrepreneurial innovation be stimulated? What sorts of barriers need to be overcome?
3. Suppose you were a slave trader during the mid-eighteenth century. How would you justify the traffic in and the institution of slavery?

TRANSITIONS

In "The Emergence of the European State System," the creation of absolutism and the alternatives to absolutism during the seventeenth and eighteenth centuries were examined, with emphasis on the process of state building.

In "The Wealth of Nations," the great economic changes of the eighteenth century and the competition for colonial empire are examined. Population growth, rising prices, and protoindustrialization helped to create general economic growth; but it was in England that a new economic system, marked by agricultural and industrial innovations, was being established. The importance of trade is emphasized by the competition between England and France for colonial empires, resulting in the Great War for Empire.

In "The Age of Enlightenment," the intellectual and cultural life of the eighteenth century will be examined.

ANSWERS

Self Test

1b; 2a; 3b; 4a; 5d; 6c; 7d; 8b; 9b; 10b; 11d; 12b; 13b; 14c; 15d; 16d; 17a; 18d; 19a

Guide to Documents

I-1b; II-1d; III-1b; IV-1c

Significant Individuals

1f; 2d; 3a; 4g; 5b; 6e; 7c

Identification

1g; 2a; 3b; 4i; 5d; 6c; 7e; 8h; 9f

NINETEEN
THE AGE OF ENLIGHTENMENT

CHAPTER HIGHLIGHTS

1. The Enlightenment was a self-conscious movement of intellectuals who elaborated on the scientific world view that originated in the seventeenth century in the hope that its new way of understanding things would make people more rational, tolerant, virtuous, and free.
2. The French philosophes led the Enlightenment, advocating science, secularism, and social utility.
3. The philosophes, suggesting either reform from above or a balance of powers, made relatively mild political and social recommendations. Rousseau stood out as a profound radical theorist in politics as well as in other fields.
4. High culture during the eighteenth century was cosmopolitan and dominated by the French. Characteristic trends were a dramatic increase in the number of publications, a flowering of academies and salons, the rise of the novel, and the development of the symphony.
5. Popular culture remained recreational, public, collective, and oral.

CHAPTER OUTLINE

I. The Enlightenment
1. The Broadening Reverberations of Science
2. Beyond Christianity
3. The Philosophes
4. Diderot and the *Encyclopedia*
5. Jean-Jacques Rousseau

II. Eighteenth-Century Elite Culture
1. Cosmopolitan High Culture
2. Publishing and Reading
3. Literature, Music, and Art

III. Popular Culture
1. Popular Literature
2. Literacy and Primary Schooling
3. Sociability and Recreation

SELF TEST

1. Who was NOT one of the three most important seventeenth century thinkers to the Enlightenment?
    a. Isaac Newton.
    b. Rene Descartes.
    c. Thomas Hobbes.
    d. John Locke.

2. Science was important to the Enlightenment as
   a. the focus of the philosophes' original thought.
   b. a source of theoretical justifications for the philosophes' programs.
   c. an inspiring example of the success of reason and experience.
   d. the major recruiting ground for new philosophes.

3. The Enlightenment eroded the authority of revealed religion through all of the following means EXCEPT
   a. toleration, which demanded the coexistence of competing revealed truths.
   b. deism, which stripped religion of its mystical and supernatural trappings.
   c. science, which proved that the universe is just a giant mechanism.
   d. criticism, which subjected Christianity to rationalist analysis.

4. One important reason the philosophes championed intellectual freedom was
   a. they felt that all points of view must be treated as equally valid.
   b. the beliefs they opposed were supported by official power.
   c. their beliefs rested not on assumptions but on facts.
   d. they wanted to mobilize the masses.

5. All of the following Enlightenment figures pioneered the social sciences EXCEPT
   a. Montesquieu, who essentially founded political science by writing a comparative study of governments.
   b. David Hume, who transformed ethics from a philosophical to a scientific field of study.
   c. Adam Smith, who advanced economics by focusing on the mechanical workings of the market.
   d. Voltaire, who moved history beyond chronicles of battles to analysis of social institutions and culture.

6. The *Encyclopedia* advanced the Enlightenment by all of the following EXCEPT
   a. conveying the philosophes' concept of useful knowledge.
   b. forcing opponents of the Enlightenment to concede by its weight of evidence.
   c. stressing the social utility of science and social science and applying rational standards to religion.
   d. overcoming official censorship in the process of satisfying the demand for works of the Enlightenment.

7. Rousseau advocated all of the following EXCEPT
   a. the founding of morals on the basis of conscience rather than reason.
   b. the creation of a new academy that would bring together scientists and humanists.
   c. education through the cultivation of natural talents rather than imposition of dry information.
   d. a social and political system in which individual desires are subordinate to group consensus.

8. The elite's cosmopolitan culture, the "republic of letters," was held together by all of the following EXCEPT
   a. a lecture circuit that paid philosophes to give public talks throughout Europe.
   b. travel, and in particular the "grand tour" of modern capitals and ancient ruins.
   c. the use of French as a common language.
   d. learned academies and salons that brought aristocrats, philosophes, and well-to-do commoners together.

9. The growth of publishing in the eighteenth century was evidenced by all of the following EXCEPT
    a. the proliferation of journals for specialized interests.
    b. the disappearance of oral traditions in popular culture.
    c. the appearance of regular newspapers.
    d. the increasing number and profitability of booksellers.

10. The novel, romantic poetry, and symphonic music had in common that they
    a. celebrated the structure that classical forms gave to human experience.
    b. emphasized emotional experience as the focus of artistic endeavor.
    c. rebelled against the constraints of classical forms by abandoning structure.
    d. opened up artistic experience to peasants and the urban working classes.

11. The social context of art was changing in that
    a. aristocrats were replacing the Church as the primary source of patronage.
    b. artists were suddenly confronted with the need to support themselves commercially.
    c. critics and public exhibitions were creating a "public sphere" of cultural discourse.
    d. artists for the first time turned their attention to ordinary people's lives.

12. Cheap books for ordinary people tended to be in all of the following genres EXCEPT
    a. almanacs.
    b. religious tracts.
    c. entertaining stories.
    d. popularizations of the Enlightenment.

13. Literacy in France was
    a. highest in the northeast.
    b. highest in the east.
    c. highest in the south.
    d. highest in the southwest.

14. Which countries did the most to promote schooling of the common people?
    a. England and France.
    b. France and Prussia.
    c. England and Austria.
    d. Austria and Prussia.

15. Popular organizations included all of the following EXCEPT
    a. journeymen's secret societies that combined social and trade-union functions.
    b. confraternities honoring saints that united established artisans and provided for a dignified funeral.
    c. mutual aid societies that collected dues and provided aid in times of illness or accident.
    d. salons in which social and cultural leaders met to exchange their views.

16. During the eighteenth century, refined cultural tastes came to be
    a. essential at all levels of society.
    b. seen by all to be hollow pretense.
    c. the sign of membership in the elite.
    d. linked to a democratic view of society.

GUIDE TO DOCUMENTS

I. Joseph II on Religious Toleration

1. What does Joseph II mean by "toleration"?
    a. Indifference about whether or not his subjects are Catholic or Protestant.
    b. Equal rights for members of all faiths.
    c. Granting of full temporal rights to Christians (and ultimately Jews).
    d. Acceptance of people of all religious orientations.

2. He justifies tolerance on all of the following bases EXCEPT
    a. the benefits to the economy and the state.
    b. avoiding the greater danger of irreligion.
    c. the private nature of religion.
    d. the benefits to civil order.

3. How are Joseph II's ideas related to the Enlightenment? Should he be considered an "enlightened" absolutist?

II. What is Enlightenment

1. According to Kant, the essence of Enlightenment is
    a. the inability to use one's own understanding.
    b. the intelligence to understand new ideas.
    c. the ability to act as guardian for lesser minds.
    d. the courage to think for yourself.

2. What, to Kant, is the relationship of enlightenment and freedom? What are the proper spheres of freedom and obedience?

III. Mary Wollstonecraft on the Education of Women

1. In what way, according to this excerpt, were women socialized?
    a. Through an education designed to develop agile bodies and sound minds.
    b. Through a false system of education in which they are led to exact respect by their abilities and virtues.
    c. Through a specious homage that causes them to degrade themselves in their attempts to gain love.
    d. Through an endeavor to acquire masculine qualities.

2. How did women respond to the role ascribed to them?
   a. Sullen acceptance.
   b. Intoxicated acceptance.
   c. Unconscious rejection.
   d. Defiant exploitation.

3. What sorts of changes in the socialization of women does Wollstonecraft suggest?

IV. Rousseau's Concept of the General Will

1. According to Rousseau, what is the "general will"?
   a. The common interest.
   b. The decision of the majority.
   c. The freedom of every individual.
   d. The duty of a subject.

2. How does Rousseau distinguish the general will from a particular will? What are the implications of this distinction?

3. According to this excerpt, can Rousseau be considered a supporter of democracy?

IDENTIFICATION

1. philosophes (fē-lo-ZOFS)
2. natural history
3. atheism
4. deism
5. social utility
6. physiocrat
7. The *Encyclopedia*
8. laissez-faire (les-ā-FAIR)
9. grand tour
10. republic of letters
11. the general will
12. freemasonry
13. provincial academies
14. bookseller
15. almanac

a. Compendium of knowledge and critique of old ideas
b. Trip to major capitals and sites of antiquities
c. Combination editor, printer, and salesperson
d. Economist who emphasized agriculture and free markets
e. Learned societies in secondary cities
f. Popular book with information on various useful topics
g. Study of geology, zoology, and botany
h. Rousseau's concept of consensus that's best for everybody
i. Enlightened intellectuals
j. Secret society promoting liberal reforms
k. Belief that God made the universe and then withdrew
l. Economic doctrine advocating free enterprise
m. The community of Enlightened intellectuals
n. Disbelief in God
o. Assessment of value based on usefulness to people

SIGNIFICANT INDIVIDUALS

1. Voltaire (vol-TAIR)
2. Denis Diderot (dē-DRŌ)
3. Jean d'Alembert (da-lan-BAIR)
4. Baron de Montesquieu (mon tes KYOO)
5. Pierre Bayle (bail)
6. G.L. Buffon (bu-FON)
7. David Hume
8. Adam Smith
9. Immanuel Kant (kahnt)
10. Marquis of Beccaria (bāk-ka-RĒ-a)
11. Edward Gibbon
12. Mary Wollstonecraft
13. Jean-Jacques Rousseau (roo-SŌ)
14. Samuel Richardson
15. Henry Fielding
16. Johathan Swift
17. William Wordsworth
18. Johann Wolfgang von Goethe (gœ-TA)
19. Jacques Louis David
20. Franz Joseph Haydn (HĪ-den)
21. Wolfgang Amadeus Mozart (MŌ-tsart)
22. Ludwig van Beethoven (BĀ-tō-ven)
23. Joseph II

a. Foremost practitioner of natural history (1707-1788)
b. Author of comic epic novels (1707-1754)
c. The last "universal man" (1749-1832)
d. Author of seminal work on constitutions, *The Spirit of the Laws,* advocating balance to restrain power (1689-1755)
e. Leader of the Enlightenment (1694-1778)
f. Leading neoclassical painter (1748-1825)
g. Economist and penal reformer (1738-1794)
h. Classical and Romantic composer (1770-1827)
i. Mathematician and leading philosophe (1717?-1783)
j. Philosopher who harmonized moral absolutes with practical reason (1724-1804)
k. Early romantic poet (1770-1850)
l. Court composer later supported by publisher (1732-1809)
m. Pioneer of novel (1689-1761)
n. Brilliant composer unable to support himself (1756-1791)
o. Editor-in-chief of the *Encyclopedia* (1713-1784)
p. Radical thinker on education and politics (1712-1778)
q. Critic who put religion to the test of reason (1647-1706)
r. Satiric author of *Gulliver's Travels* (1667-1745)
s. Tolerant Habsburg monarch (r.1765-1790)
t. Early advocate of women's rights (1759-1797)
u. Philosopher who analyzed good and evil in pragmatic terms (1711-1776)
v. Historian who blamed the decline and fall of the Roman Empire on Christianity (1737-1794)
w. Economist, author of *Wealth of Nations* (1723-1790)

PROBLEMS FOR ANALYSIS

I. The Enlightenment

1. Compare the Enlightenment with the scientific revolution. What are their main points of similarity and difference?
2. "Voltaire, more than any other man, represents the philosophes and what they stood for." Do you agree? Why?
3. "The *Encyclopedia,* more than any other book, represents the meaning of the Enlightenment." Do you agree? Why?

II. Eighteenth-Century Elite Culture

1. Compare high and popular culture of the eighteenth century. How are distinctions between the two revealed in differing types of publications and the role of the bookseller?
2. It has been argued that the major cultural accomplishments of the eighteenth century were development of the novel and of the symphony. Do you agree? Why?
3. What were the roles of women in eighteenth-century cultural life?

III. Popular Culture

1. What role did popular culture play in the lives of ordinary people?
2. How does the nature and extent of public schooling reflect broad aspects of eighteenth-century society and culture?

SPECULATIONS

1. Suppose you were a philosophe serving as an influential adviser to a monarch. What sort of policies would you recommend? Why?
2. Suppose there were a debate between Voltaire and Rousseau. Describe the points each would be likely to make and how they would respond to each other.
3. Why, do you think, was the Enlightenment and much of the culture of the age so dominated by the French?
4. Do you think we remain a society dominated by the ideas of the Enlightenment?

TRANSITIONS

In "The Wealth of Nations," the economic changes sweeping England and some other areas during the eighteenth century were analyzed, as was the colonial rivalry between England and France.

In "The Age of Enlightenment," the intellectual and cultural currents of the eighteenth century are examined. This was a culturally and intellectually transitional time in which a body of Enlightenment thinkers, with a self-conscious mission of enlightening people, attempted to reform traditional ideas and attitudes along the lines initiated by the scientific and philosophical innovations of the seventeenth century. In doing this, they helped put the status quo on the defensive. New developments marked elite and popular culture during the period.

In "The French Revolution," the revolution and reforms that swept across France toward the end of the century, affecting much of Europe in the process, will be analyzed.

ANSWERS

Self Test

1c; 2c; 3c; 4b; 5b; 6b; 7b; 8a; 9b; 10b; 11c; 12d; 13a; 14d; 15d; 16c

Guide to Documents

I-1c; I-2c; II-1d; III-1c; III-2b; IV-1a

Significant Individuals

1e; 2o; 3i; 4d; 5q; 6a; 7u; 8w; 9j; 10g; 11v; 12t; 13p; 14m; 15b; 16r; 17k; 18c; 19f; 20l; 21n; 22h; 23s

Identification

1i; 2g; 3n; 4k; 5o; 6d; 7a; 8l; 9b; 10m; 11h; 12j; 13e; 14c; 15f

SECTION SUMMARY
THE EARLY MODERN PERIOD 1560–1789
CHAPTERS 15–19

CHRONOLOGICAL DIAGRAM

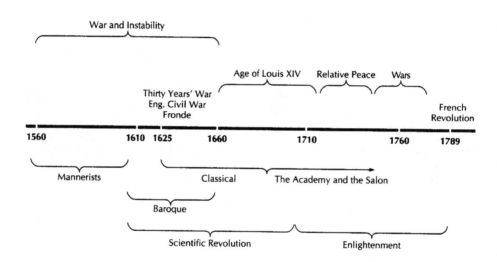

MAP EXERCISE

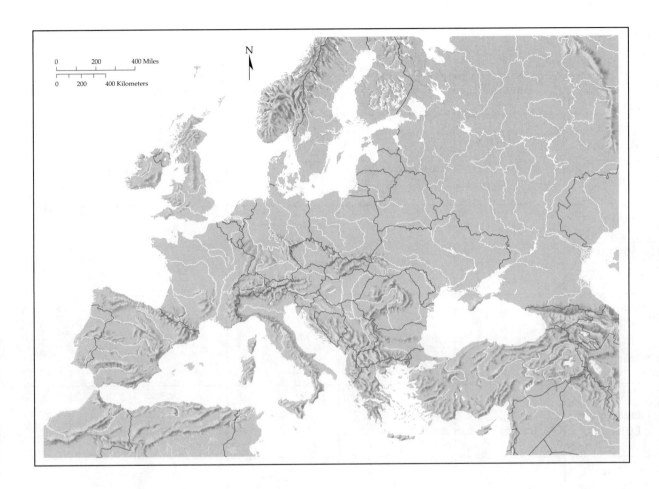

1. Indicate the main political divisions of Europe in 1648.
2. Indicate who gained most from the 1648 Peace of Westphalia and who lost most.
3. Indicate which European nations would be on the rise during the eighteenth century.

BOX CHART

Reproduce the Box Chart in a larger format in your notebook or on a separate sheet of paper. For a fuller explanation of the themes and how best to find material, see Introduction.

Chart 1:

It is suggested that you devote one page for each column (i.e., chart all seven themes for each country on a page).

Countries / Themes	England	France	Netherlands	Austria	Prussia
Social Structure: Groups in Society					
Politics: Events and Structures					
Economics: Production and Distribution					
Family Gender Roles Daily Life					
War: Relationship to larger society					
Religion: Beliefs, Communities, Conflicts					
Cultural Expression: Formal and Popular					

Chart 2:

It is suggested that you devote one page for each row (i.e., chart each of the seven themes on a separate page; you may want to use blank or graph paper and turn the page sideways). Note major national differences, but keep the focus on development in Europe as a whole.

Periods Themes	1560	1648	1715	1789
Social Structure: Groups in Society				
Politics: Events and Structures				
Economics: Production and Distribution				
Family Gender Roles Daily Life				
War: Relationship to larger society				
Religion: Beliefs, Communities, Conflicts				
Cultural Expression: Formal and Popular				

CULTURAL STYLES

1. Examine the pictures on page 512, 517, 518, 522, and 542. What do they suggest about the conditions of life during the period?

2. Compare the picture on page 515 with the ones on 568, 569, and 570? How do they illustrate the differences between the Mannerist and the Baroque styles?

3. Compare the pictures on pages 575 and 588 with the ones on pages 568, 569, and 570. How is the Classical style different from the Baroque?

4. Compare the view of the Vatican on page 572 with that of Versailles on page 591. In what ways are the two buildings similar? In what ways different? What do the similarities and differences suggest about the purposes for which they were built. What do they suggest about the changing roles of Church and state?

5. Contrast the pictures on pages 593, 615, and 687 with those on pages 568, 569, and 570. How are they different in style? What are the differences in their subject matters? What do those differences suggest about the larger differences in European life in the two periods?

6. Compare the pictures on pages 649 and 657 with the one on page 651. How do they reflect the differences between life in the core areas of the Western economy and life on its periphery?

7. Examine the pictures on pages 632, 637, 641, 651, and 653. In what ways do they reflect life during the period they were made? Contrast them with the pictures you looked at in question 1. What do the differences in their subject matters suggest about the changes in Europe during the early modern period?

8. Contrast the pictures on pages 632, 637, 641, 651, and 653 with those on pages 593, and 687. In what ways are the activities depicted complementary? In what ways are they at odds?

9. Compare the picture on page 687 with those on page 688 and 689. What are the different sensibilities being expressed? Is it fair to say that the latter two are a challenge to the former?

TWENTY
THE FRENCH REVOLUTION

CHAPTER HIGHLIGHTS

1. Movements for reform, from above as well as from privileged and unprivileged estates, grew during the eighteenth century, provoking crises and political unrest.
2. The French Revolution—supporting ideals of liberty, equality, and fraternity—broke out in 1789, leading eventually to the fall of the French monarchy and a series of sweeping reforms in many areas of civic and social life.
3. A second revolution occurred in 1792, leading France into a more radical phase dominated by Jacobins on the Committee of Public Safety, by the sans-culottes, and by the Terror.

CHAPTER OUTLINE

I. Reform and Political Crisis
1. "Enlightened Absolutism" in Central and Eastern Europe
2. Joseph II and the Limits of Absolutism
3. Constitutional Crises in the West
4. Upheavals in the British Empire

II. 1789: The French Revolution
1. Origins of the Revolution
2. Fiscal Crisis and Political Deadlock
3. From the Estates General to the National Assembly
4. The Convergence of Revolutions

III. The Reconstruction of France
1. The Declaration of the Rights of Man and Citizen
2. The New Constitution
3. The Revolution and the Church
4. Counterrevolution, Radicalism, and War

IV. The Second Revolution
1. The National Convention
2. The Revolutionary Crisis
3. The Jacobin Dictatorship
4. The Sans-Culottes: Revolution from Below
5. The Revolutionary Wars

SELF TEST

1. Joseph II instituted all of the following "enlightened" reforms EXCEPT

   a. free expression, religious toleration, and greater state control of the Church.

   b. the end of judicial torture and class distinctions in the administration of justice.

   c. convocation of the estates as a parliamentary forum for political discussion.

   d. abolition of serfdom and creation of a freeholding peasantry.

2. "Enlightened absolutism" failed to achieve many lasting reforms for all of the following reasons EXCEPT

   a. lack of genuine commitment on the part of some of the monarchs.

   b. a preference for administrative reforms rather than empowerment of citizens.

   c. entrenched opposition from the aristocracy, the clergy, and privileged, or just conservative, commoners.

   d. the efforts weakened the states and made them vulnerable to conquest by others.

3. The pre-Revolutionary constitutional crises in western Europe were fueled by all of the following EXCEPT

   a. attempts by the aristocracy to improve their position visa-vie the monarchy.

   b. general correspondence between the views of the clergy and those of the aristocracy.

   c. democratic agitation by unprivileged commoners against both monarchy and aristocracy.

   d. overtures by the monarchs to the commoners to form a common front against the aristocracy.

4. Upheavals in the British Empire included all of the following EXCEPT

   a. radical agitation by merchants and gentry against the crown and aristocracy.

   b. the struggle between John Wilkes and the crown over his vicious criticisms of the government.

   c. agitation by groups of solid citizens for parliamentary reform.

   d. opposition and ultimately rebellion in Britain's North American colonies.

5. Political agitation in America differed fundamentally from that in Europe because the Americans

   a. appealed to traditional rights of all British and theories of popular sovereignty and natural rights.

   b. focused on practical issues like opposition to taxation rather than abstract principles of liberty.

   c. grounded their protests in the body of privileges that the monarchy violated with its new tax plans.

   d. organized their protests on the basis of interest groups: landowners, merchants, and artisans.

6. The American Revolution was pathbreaking for all of the following reasons EXCEPT

   a. it maintained an alliance between social strata based on a shared commitment to legal equality.

   b. it created the first modern government based on participation and consent of the citizens.

   c. it established political rights and legal equality for all adult men.

   d. it was the first successful rebellion by overseas colonies against their European masters.

7. The Revolution of 1789 was caused by all of the following EXCEPT
    a. new theories of politics emphasizing rationality and popular sovereignty over tradition and divine right.
    b. growing assertiveness on the part of the aristocracy that made it ready to exploit the king's fiscal woes.
    c. the desire of the bourgeoisie to win political control of the nation to match its socioeconomic dominance.
    d. popular discontent caused by overpopulation, crop failures, grain shortages, and high levels of vagrancy.

8. The core of the monarchy's fiscal problem was its
    a. historic inability to tax privileged groups.
    b. lavish spending on court festivities and frivolities.
    c. refusal to consider significant reform measures to tap the nation's wealth.
    d. inability to find competent ministers to put reform measures into place.

9. The critical question as the Estates General met was whether it would meet as one or three chambers because
    a. the Third Estate had twice as many representatives, and so could dominate a combined chamber.
    b. the nobles and clergy plus conservative commoners would dominate a combined chamber.
    c. the double vote given the Third Estate meant that three separate chambers would constantly deadlock.
    d. the liberal clergy could ally the First Estate with the Third to achieve a majority of two of three chambers.

10. The Estates General legally became the National Assembly when
    a. Louis XVI and Necker met with it on May 5 and established its voting procedures.
    b. the Third Estate invited the others to join it and proclaimed itself the National Assembly..
    c. the king, faced with the Third Estate's refusal to meet separately, ordered the other estates to join it.
    d. the people of Paris stormed the Bastille and took over the municipal government.

11. The different revolutions that converged during the summer of 1789 included all of the following EXCEPT
    a. the Parisian insurrection of July 14 that forestalled a military coup by royalist forces.
    b. the "Great Fear," in which peasants across the country rebelled against their seigniorial landlords.
    c. the August 4 Decree abolishing feudalism and the Declaration of the Rights of Man setting the new order.
    d. the storming of the Tuileries and the suspension of the king by the Legislative Assembly.

12. Between 1789 and 1791 the National Assembly accomplished all of the following EXCEPT
    a. shifting power to citizens with property and creating a limited monarchy and powerful legislature.
    b. extending civil and political equality to Jews and blacks in France's Caribbean colonies.
    c. reforming and standardizing local administration and replacing the old judicial system.
    d. applying principles of laissez-faire to economics.

13. The National Assembly's religious policy included all of the following EXCEPT
    a. confiscating ecclesiastical property, issuing notes backed by its value, and auctioning off the land.
    b. reforming the administration of the Catholic Church in France.
    c. encouraging the clergy to renounce their vocations and marry.
    d. requiring the clergy to take an oath of loyalty to the constitution.

14. The effect of the National Assembly's religious policy was to
    a. lay a lasting foundation for the government's finances.
    b. alienate most of the clergy and many devout French people.
    c. undermine the long-term viability of Catholicism in France.
    d. create the basis of a lasting partnership between church and state.

15. The monarchy fell in August, 1792 because of all of the following EXCEPT
    a. the king's long-standing and increasingly apparent opposition to the revolution.
    b. the outbreak of war against Austria, Prussia, and the counter-revolutionary émigrés.
    c. the king's veto of measures to suppress dissidents and mobilize the national guard around Paris.
    d. the royalist coup attempt by political prisoners held in Parisian jails supported by the Duke of Brunswick.

16. The three main parties in the National Convention included all of the following EXCEPT
    a. the Girondins, who advocated provincial liberty and laissez-faire economics.
    b. the Mountain, the more radical faction, which demanded bold measures to protect the revolution.
    c. the Plain, the majority of delegates, who were uncertain which path to follow.
    d. the Sans-culottes, who denounced even the Mountain as too moderate.

17. The Convention faced violent opposition from all of the following EXCEPT
    a. anti-revolutionary peasants, priests, émigrés, royalists, and moderates in the west and south.
    b. the Paris Commune, which insisted that it actually embodied the will of the nation.
    c. Parisian sans-culottes who wanted radical measures like price controls and execution of speculators.
    d. an enlarged foreign coalition including Austria, Prussia, Spain, Piedmont, and Britain.

18. After the purge of the Girondins, the Jacobin Dictatorship was characterized by all of the following EXCEPT
    a. implementation of the constitution drafted by the Convention to deal with the emergency situation.
    b. laws imposing price controls and authorizing arrest of suspected traitors.
    c. rule by the "Committee of Public Safety" led by Robespierre.
    d. a "Reign of Terror" that killed anti-revolutionaries, political opponents, and citizens thought suspicious.

19. The sans-culottes were politically mobilized urban common people who favored all of the following EXCEPT
    a. price controls and punishment of profiteers.
    b. antiaristocratic styles of dress, manners, and morals.
    c. decentralized, direct democracy.
    d. land reform to increase the holdings of small farmers.

20. The Revolutionary war effort was successful for all of the following reasons EXCEPT
    a. the enthusiasm of soldiers fighting to defend their newly won freedoms.
    b. the ability of the state to mobilize men and material on a far greater scale than before.
    c. the new tactics using massive attack columns in place of the well-drilled lines of the old army.
    d. the massive uprisings of common people in bordering countries and the enemy nations.

GUIDE TO DOCUMENTS

I. Two Views of the Rights of Man

1. The two documents agree that the rights of man are grounded in
    a. the estates of the state, which define each individual's specific rights and responsibilities.
    b. liberty, which is a man's right to do as he wants so long as it does not harm anyone.
    c. the sovereignty of the nation, which must expressly confer authority for someone to act in its name.
    d. the right to representation in the formulation of the laws and the levying of taxes.

2. The principal difference between the two documents is that the "Declaration of the Rights of Man" asserts that
    a. all men are equal in their rights and before the law.
    b. liberty is the right to do whatever does not harm another.
    c. sovereignty rests essentially with the nation.
    d. taxation should be apportioned equally according to each citizen's ability to pay.

3. What explains the greater appeal of the ideas in the "Declaration of the Rights of Man and Citizen" over those expressed by the "Prussian General Code?"

II. Robespierre's Justification of the Terror

1. The essence of Robespierre's argument is that
    a. violence is morally purifying.
    b. the end justifies the means.
    c. violence in the defense of liberty is not violence.
    d. internal enemies are more dangerous than foreign ones.

2. What does this excerpt reveal about the nature of the most radical phase of the Revolution?

III. A Portrait of the Parisian Sans-Culotte

1. Which is NOT an attribute of the sans-culottes that helps explain their political agenda?
    a. They were working people, so they wanted a government that protected the common peoples' interests.
    b. They were upright people, so they favored a government that enforced an austere morality.
    c. They were self-sacrificing people, so they were ready to drop everything to fight, and die, for the Republic.
    d. They were moderate people, so they opposed mob rule and passionate political posturing.

2. What reveals this as a sympathetic picture of the sans-culottes? What might a more antagonistic picture of the sans-culottes emphasize?

SIGNIFICANT INDIVIDUALS

1. Catherine II
2. George III
3. Joseph II
4. Louis XVI
5. John Wilkes
6. Jacques Turgot (tur-GŌ)
7. Jacques Necker
8. Charles Alexander de Calonne
9. Loménie de Brienne (brē-EN)
10. Emmanuel Sieyès (syā-YES)
11. Jean Paul Marat (ma-RA)
12. Georges Danton (dahn-TON)
13. Maximilien Robespierre (rōbz-PYAIR)
14. Réné Hébert (er-BAIR)

a. Enlightened Habsburg Emperor (r.1765-1790)
b. Popular British agitator and politician (1727-1797)
c. Early leader of the Third Estate (1748-1836)
d. Controller-general who foresaw bankruptcy (r.1783-1787)
e. Not-so-enlightened Empress of Russia (r.1762-1796)
f. Archbishop and controller-general (r.1787-1788)
g. Radical journalist and Jacobin deputy (1743-1793)
h. Radical executed because not radical enough (1759-1794)
i. Radical executed because too radical (1755-1794)
j. English King who tried party politics (r.1760-1820)
k. Controller-general who advocated reforms (r.1774-1776)
l. Leader of Jacobins during Reign of Terror (1758-1794)
m. Swiss financier and French controller-general (1776-1781, 1788-1790)
n. King of France who lost his head (r.1774-1793)

CHRONOLOGICAL DIAGRAM

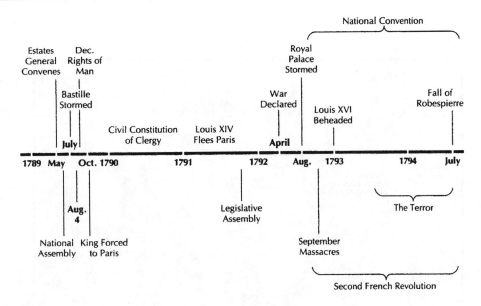

IDENTIFICATION

1. the Third Estate
2. Declaration of the Rights of Man
3. *assignats* (a-sē-NYA)
4. *cahiers* (kah-YĀ)
5. the "Great Fear"
6. Civil Constitution of the Clergy
7. Jacobin Club (JAK-o-bin)
8. Girondists (ji-ROND-ists)
9. the Mountain
10. Vendée rebellion (vahn-dā)
11. sans-culottes (sahn-kuh-LOT)
12. Law of Suspects
13. Committee of Public Safety
14. *levée en masse* (levā ahn mas)

a. Organization of radicals centered in Paris
b. Proclamation of French Revolutionary principles
c. Lower-class Parisian radicals
d. Radical faction in the National Assembly
e. Statue permitting incarceration of suspected traitors
f. Mobilization of unmarried young men for military service
g. Everybody else
h. Grievance petitions drafted prior to Estates General
i. Executive body overseeing the Reign of Terror
j. Peasant uprising accompanying initial revolution
k. Revolutionary currency secured by Church property
l. Faction for provincial liberty and laissez-faire economics
m. Anti-revolutionary uprising in western France
n. Law establishing state's control over the Church

MAP EXERCISES

1. Indicate those areas of Europe experiencing substantial revolutionary activity during the last quarter of the eighteenth century.

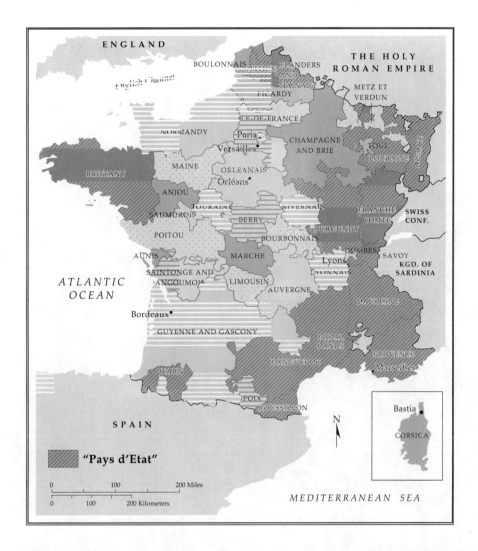

1. The map on this page shows French provinces and regions before 1789. The map on the next page shows France's revolutionary departments after 1789. Together, what do these maps reveal about the way in which political reform was carried out in France? What do these maps reveal about the political and administrative significance of the French Revolution for France?

1. This map shows Europe in 1799. What does it reveal about the growth of France and the influence of the French Revolution by 1799?

PROBLEMS FOR ANALYSIS

I. Reform and Political Crisis

1. The theory that certain eighteenth-century monarchs distinguished themselves as "enlightened absolutists" has been called a distortion of history. Yet it seems that Joseph II, Frederick the Great, and perhaps even Catherine II displayed some "enlightened" qualities. How do you explain this apparent contradiction?

2. What common characteristics connect the various "constitutional crises" in the West? How might these crises be explained?

II. 1789: The French Revolution

1. In what ways might the causes for the French Revolution be explained? Which explanation do you think is most convincing?

2. Focusing only on the fiscal crises, do you think it is accurate to argue that the Revolution was caused by the conflict between the king and the aristocracy over taxation policies? Why?

III. The Reconstruction of France

1. How was France restructured during the first three years of the French Revolution?
2. Who benefited most from this restructuring? Who lost most?

IV. The Second Revolution

1. What problems faced the national Convention? What policies did it pursue to solve those problems? How successful do you think the government was in solving those problems?

2. Compare the role of the sans-culottes with that of the peasantry in the revolutionary developments between 1789 and 1795.

SPECULATIONS

1. Do you think it would have been possible for a truly "enlightened" French monarch to institute many of the reforms of the French Revolution and thereby avoid the violence of a revolution? Explain.

2. Once the Revolution had started in 1789, what policies should Louis XVI have followed to gain control over it? Explain.

3. Develop an argument justifying the Terror, in light of the circumstances and the goals of governmental leaders. Counter that argument with one which holds that the Terror was unjustified, even in light of the circumstances and the goals of governmental leaders. Which argument do you think is strongest?

TRANSITIONS

In "The Age of Enlightenment," the intellectual and cultural currents of the eighteenth century were analyzed, with particular focus upon the philosophes in France.

In "The French Revolution," the political transformations occurring toward the end of the century are examined. The French Revolution asserted the claim to power of the unprivileged classes and ended both the traditional social order and the monarchy. In 1792 a second revolution brought to power the radical Jacobins, supported by the sans-culottes.

In "The Age of Napoleon," the last stage of the Revolution and the ensuing rule of Napoleon will be examined.

ANSWERS

Self Test

1c; 2d; 3d; 4a; 5a; 6c; 7c; 8a; 9a; 10c; 11d; 12b; 13c; 14b; 15d; 16d; 17b; 18a; 19d; 20d

Guide to Documents

I-1b; I-2a; II-1b; III-1d

Significant Individuals

1e; 2j; 3a; 4n; 5b; 6k; 7m; 8d; 9f; 10c; 11g; 12h; 13l; 14i

Identification

1g; 2b; 3k; 4h; 5j; 6n; 7a; 8l; 9d; 10m; 11c; 12e; 13i; 14f

TWENTY-ONE
THE AGE OF NAPOLEON

CHAPTER HIGHLIGHTS

1. The government, having surmounted the internal and external crises, turned against the sans-culottes—its earlier supporters. This development paved the way for the Thermidorian reaction, the establishment of the more moderate Directory, and the eventual rise of Napoleon Bonaparte.

2. Napoleon consolidated and firmly established many of the institutional gains of 1789 while rejecting more radical revolutionary policies. He then expanded the French imperium over the continent; but he finally extended himself too far and was defeated in battle by a coalition of opposing powers in 1814.

CHAPTER OUTLINE

I. From Robespierre to Bonaparte
1. The Thermidorian Reaction (1794–1795)
2. The Directory (1795–1799)
3. The Rise of Bonaparte
4. The Brumaire Coup

II. The Napoleonic Settlement in France
1. The Napoleonic Style
2. Political and Religious Settlements
3. The Era of the Notables

III. Napoleonic Hegemony in Europe
1. Military Supremacy and the Reorganization of Europe
2. Naval War with Britain
3. The Napoleonic Conscription Machine

IV. Opposition to Napoleon
1. The "Spanish Ulcer"
2. The Russian Debacle
3. German Resistance and the Last Coalition
4. The Napoleonic Legend

SELF TEST

1. The Thermidorian reaction involved all of the following EXCEPT
   a. the fall from power and execution of Robespierre.
   b. the end of the Terror and the "white terror" against the radicals.
   c. the end of the Revolution and restoration of the monarchy.
   d. the reversal of egalitarian political policies and social mores.

2. The Directory tried to ground itself in the support of the political
   a. right, the ultraroyalists and moderate royalists hoping for restoration.
   b. center, the well-to-do citizens who wanted to conserve the accomplishments of 1789 to 1791.
   c. left, the neo-Jacobins who wanted to preserve the gains of 1793 without the Terror.
   d. far left, Gracchus Babeuf's collectivist followers who hoped to move the Revolution even farther.

3. Napoleon's political stature grew because of all of the following EXCEPT
   a. his brilliant military and diplomatic successes in northern Italy.
   b. the dependence of the unpopular Directory on military success.
   c. his personal charisma and talent in manipulating people.
   d. the success of his military expedition to Egypt.

4. The "revisionists" enlisted Napoleon in the conspiracy that ended in the Brumaire Coup because
   a. they recognized his genius and wanted to install him as dictator.
   b. they hoped to use him as a figurehead since he was very popular.
   c. they needed the full power of the army to avoid renewed civil war.
   d. they feared that if they did not include him he would turn on them.

5. The principles of the Revolution that Napoleon preserved included all of the following EXCEPT
   a. disdain for the unjust and cumbersome institutions of Bourbon absolutism.
   b. rejection of seigniorialism and traditional aristocratic privileges.
   c. commitment to popular sovereignty and representative institutions.
   d. promotion of rational institutions and equality of opportunity.

6. The institutional initiatives that Napoleon implemented included all of the following EXCEPT
   a. concentration of power into his own hands as First Consul, then Consul-for-Life, and finally Emperor.
   b. subordination of local government to central control through appointed prefects, sub-prefects, and mayors.
   c. conclusion of a Concordat with the Church ending some revolutionary measures but keeping state control.
   d. restoration of the rule of law and the abolition of the institutions of police state set up by the Directory.

7. Napoleon favored the rule of "notables" (talented and wealthy men) through all of the following EXCEPT
    a. appointing them to positions power and honoring them with distinctions.
    b. allowing them to elect the experts who staffed the advisory Council of State.
    c. creating a system of elite secondary schools to prepare future officials, engineers, and officers.
    d. codifying patriarchy and the rights of property in the Napoleonic Code of civil law.

8. The Napoleonic Code established all of the following EXCEPT
    a. legal equality and freedom of worship.
    b. modern contractual notions of property.
    c. the right to organize unions and strike.
    d. the right to choose one's profession.

9. Napoleon's victories over Austria in 1800 and 1805 established French dominance in
    a. southern Germany and Italy.
    b. Italy and Spain.
    c. northern Germany and Poland.
    d. the Netherlands and Britain.

10. Napoleon's victory over Prussia in 1806 established French dominance in
    a. northern Germany and Poland.
    b. northern Germany and Russia.
    c. southern Germany and Italy.
    d. Italy and Spain.

11. Napoleon's campaigns against Russia in 1807 to 1808 resulted in
    a. a crushing victory that gave France control of the Russian Empire.
    b. a narrow victory that gave France control over Russia's western provinces.
    c. an agreement that divided Europe into French and Russian spheres of influence.
    d. a devastating defeat that was the beginning of the end for Napoleon.

12. Napoleon created the Continental System to cut off British trade with Europe because
    a. Admiral Nelson's defeat of the French and Spanish navies at Trafalgar made an invasion impossible.
    b. he calculated that it was the most cost-effective way to bring the British Empire to its knees.
    c. he realized that protecting European industry would make it stronger than Britain's in the long run.
    d. Alexander III persuaded him that their joint dominance of Europe would be secured by economic links.

13. The effects of Napoleon's economic war with Britain included all of the following EXCEPT
    a. curtailing British business and sparking war-weariness and labor unrest.
    b. provoking a British counter-blockade that cut Europe off from overseas.
    c. alienating Napoleon's European subjects, who bore the brunt of the burden.
    d. forcing Britain to come to terms in order to stave off an economic collapse.

14. Napoleon was able to insure a flow of new recruits to his army by all of the following means EXCEPT
    a. assigning quotas to each department, which were then fulfilled by a lottery among all fit 19 year olds.
    b. sending troops to sweep through areas with high levels of draft evasion to punish evaders' families.
    c. calling up younger men and older men who had high lottery numbers during emergencies.
    d. forcing satellite countries and allied nations to send men to join the French army during 1814.

15. French rule in Spain was resisted by all of the following EXCEPT
    a. the Spanish army.
    b. the Bourbon dynasty.
    c. guerrilla bands.
    d. an English army.

16. Of the 600,000 soldiers Napoleon led into Russia, he lost
    a. less than 50,000.
    b. around 100,000.
    c. at least 250,000.
    d. MORE THAN 500,000!

17. Prussia built up its strength after the defeat of 1806 by all of the following EXCEPT
    a. opening more high positions to non-nobles and reducing some noble privileges.
    b. creating an army reserve to increase the number of men who could be mobilized.
    c. establishing the concepts of citizenship, representation, and constitutional rule.
    d. promoting the idea of a war of national liberation "With God for King and Fatherland."

18. Napoleon's legend rests on all of the following foundations EXCEPT
    a. his dramatic and manifold achievements in war and government.
    b. his self-conscious attempts to style his own image during his rule and in his memoirs.
    c. the congruence of his life with the Romantic concepts of genius and heroism.
    d. the fact that he was a nice guy.

GUIDE TO DOCUMENTS

I. Family and Gender Roles Under the Napoleonic Civil Code

1. The Napoleonic Code subordinated wives to husbands in all of the following ways EXCEPT
    a. the wife owed the husband obedience and had to live wherever he chose to reside.
    b. the wife could not represent herself in court or conclude contracts without his consent.
    c. the wife could not sue for divorce on the grounds of adultery unless the other woman moved in.
    d. the wife could not sue for divorce on the basis of outrageous conduct, ill-usage, or grievous injuries.

2. How might these provisions be important for the structure of the family and the roles to be played by various members of the family during the nineteenth century?

II. Spanish Liberals Draft a Constitution, 1812

1. This constitution echoes the Declaration of the Rights of Man in all the following ways EXCEPT

    a. the assertion that sovereignty rests primarily in the nation.

    b. the establishment of representative institutions to give citizens a voice in the formation of laws.

    c. the principles that taxes should be approved by the people and apportioned by ability to pay.

    d. the renunciation of privilege on the basis of the equality of all citizens.

2. In what ways does this constitution reflect the growing importance of nationalism?

III. Napoleon Justifies Himself in 1815

1. How does Napoleon justify his actions?

    a. The ends he pursued justified the means he used.

    b. They were essentially reactive; circumstances drew him toward goals he never desired.

    c. The negative ones were responses to necessity, while the positive were the fulfillment of human potential.

    d. He did nothing disreputable; the allegations against him were groundless misrepresentations and lies.

2. How might one refute Napoleon's points? How might such a document form a basis for the Napoleonic legend?

SIGNIFICANT INDIVIDUALS

1. Maximilien Robespierre (rōbz-PYAIR)
2. Emmanuel Sieyès (syā-YES)
3. Napoleon Bonaparte (BŌN-a-part)
4. Alexander I
5. Francis I
6. Frederick William III
7. Horatio Nelson
8. Duke of Wellington
9. Simon Bolivar (sē-MŌN bō-LĒ-var)

a. General, politician, statesman, emperor, exile, and legend

b. King of Prussia who lost and then won (r.1797-1840)

c. Leader whose fall was the turn of the tide in the Revolution (1758-1794)

d. Tsar who divided Europe with Napoleon (r.1801-1825)

e. British general who helped beat Napoleon (1769-1852)

f. Napoleon's foe and father-in-law (r.1804-1835)

g. British admiral who frustrated Napoleon (1758-1804)

h. Early revolutionary also active in Directory (1748-1836)

i. "The Liberator" of Spanish America (1783-1830)

CHRONOLOGICAL DIAGRAM

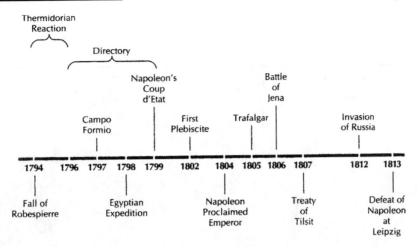

IDENTIFICATION

1. Thermidorian reaction
2. The Directory
3. 18 Brumaire
4. prefect
5. Council of State
6. Concordat of 1801
7. Order of the Legion of Honor
8. Napoleonic Code
9. "conscription machine"
10. Trafalgar
11. Continental System
12. the "Spanish Ulcer"
13. Dos de Mayo
14. Cortes of Càdiz (KOR-tez of KAH-diz)
15. Creoles (KRĒ-ōls)
16. the Napoleonic legend

a. Lasting system of laws
b. The political and social backlash against Jacobin rule
c. Liberal Spanish assembly
d. Honorary society of notables who served with distinction
e. Republican government with five man executive
f. Attempt to cut England off from Europe economically
g. Napoleon's system for drafting men into the army
h. Descendants of Spanish settlers born in the New World
i. Napoleon's reconciliation with the Catholic Church
j. Official in charge of French department
k. Date and name of Napoleon's coup
l. Riot and reprisals that opened the Spanish war
m. Protracted guerrilla war that drained French resources
n. Naval battle that ended the possibility of a French invasion of England
o. Advisory body of appointed experts
p. Romantic image of Napoleon's person and reign

MAP EXERCISES

1. Indicate the outlines of France in 1798.
2. Indicate and label the French Empire, the satellite kingdoms, the nominal allies of Napoleon, and those hostile to Napoleon in 1810.

1. Show the approximate dates of rebellion and independence of Spanish colonies.

PROBLEMS FOR ANALYSIS

I. From Robespierre to Bonaparte

1. In what ways was the Thermidorian reaction a return to a more moderate phase of the Revolution?
2. How do you explain Napoleon's rise to power? What role did luck, public relations, and conspiracy play in his ascendancy?

II. The Napoleonic Settlement in France

1. Analyze Napoleon's domestic policies. Did he repudiate the Revolution, or did he solidify and institutionalize it?

III. Napoleonic Hegemony in Europe

1. Compare Napoleon's successes and failures in war and in diplomacy.
2. What were the consequences of France's conquest of the continent? What role did the Continental System play in this outcome?

IV. Opposition to Napoleon

1. How do you explain the rise of successful opposition to Napoleon?
2. Compare the legend of Napoleon with the reality of Napoleon.

SPECULATIONS

1. Was Napoleon simply the right man at the right time in the right place, or did he create his own opportunities and exploit them through the power of his personality? Explain.
2. In what ways should Napoleon be considered an enlightened despot? In what ways should he be considered a preserver of the Revolution?

TRANSITIONS

In "The French Revolution," the Revolutions of 1789 and 1792 were analyzed. The chapter ended with events in France up to 1794.

In "The Age of Napoleon," the story is continued. The end of the crisis brought some moderation after 1794, but instability ended only with the rise to power of Napoleon Bonaparte. His attempt to reorganize Europe under French hegemony ultimately led to his fall, but Europe's experience of revolution and Napoleonic transformation initiated the era of modern political and social conflicts.

In "Foundations of the Nineteenth Century: Politics and Social Change," the effort to restore equilibrium in Europe after the Congress of Vienna and the course of industrialization will be examined.

ANSWERS

Self Test

1c; 2b; 3d; 4b; 5c; 6d; 7b; 8c; 9a; 10a; 11c; 12a; 13d; 14d; 15b; 16d; 17c; 18d

Guide to Documents

I-1d; II-1d; III-1c

Significant Individuals

1c; 2h; 3a; 4d; 5f; 6b; 7g; 8e; 9i

Identification

1b; 2e; 3k; 4j; 5o; 6i; 7d; 8a; 9g; 10n; 11f; 12m; 13l; 14c; 15h; 16p

TWENTY-TWO
FOUNDATIONS OF THE NINETEENTH CENTURY: POLITICS AND SOCIAL CHANGE

CHAPTER HIGHLIGHTS

1. At the Congress of Vienna and during the restoration, conservatives tried to reestablish stability and social order through domestic and international political arrangements.
2. During the first half of the nineteenth century, industrialization transformed modes of production, first in Great Britain and then on the continent.
3. A variety of social changes, including class development and social differentiation, accompanied and interacted with industrialization.

CHAPTER OUTLINE

I. The Politics of Order
1. The Congress of Vienna
2. The Pillars of the Restoration: Russia, Austria, Prussia
3. The Test of Restoration: Spain, Italy, and France

II. The Progress of Industrialization
1. The Technology to Support Machines
2. The Economic Effects of Revolution and War
3. Patterns of Industrialization

III. The Social Effects
1. The Division of Labor
2. The Family
3. The Standard of Living

SELF TEST

1. The five Great Powers at the Congress of Vienna were
    a. France, England, Austria, Prussia, and Russia.
    b. England, Austria, Spain, Prussia, and Russia.
    c. Russia, Austria, Sweden, England, and France.
    d. Austria, Prussia, Russia, England, and Turkey.

2. Each of the following territorial arrangements were designed to create barriers to French aggression EXCEPT
   a. Holland got Belgium to create a strong power in the northeast.
   b. Prussia got territories in the Rhineland to increase its position to the east.
   c. Austria got territory and influence in Italy to put it firmly to the southeast.
   d. Spain got Portugal to create a strong power in the southwest.

3. The Congress of Vienna started all of the following to promote long term stability EXCEPT
   a. The Concert of Europe.
   b. The Holy Alliance.
   c. the *Zollverein*.
   d. modern diplomacy.

4. Nicholas I particularly focused his attention on all of the following to make them pillars of his rule EXCEPT
   a. the Orthodox Church.
   b. the army.
   c. the secret police.
   d. the bureaucracy.

5. The Habsburgs had to contend with all of the following sources of opposition EXCEPT
   a. the aristocracy.
   b. lawyers and merchants.
   c. their own bureaucracy.
   d. Hungarian nationalists.

6. Prussia's major foreign policy accomplishment in the two decades after the Congress of Vienna was
   a. dominating the German Confederation.
   b. creating a German customs union.
   c. leading the suppression of academic radicalism.
   d. championing Polish rights against Russian oppression.

7. The two decades after the Congress of Vienna saw liberal uprisings in all of the following EXCEPT
   a. Spain in 1820, when a constitutional regime was established until French intervention suppressed it.
   b. Italy in 1820, when Austrian armies intervened to suppress constitutionalism in Naples and Piedmont.
   c. Austria in 1825, when Hungarian rebels forced the Emperor to remove Metternich from office.
   d. France in 1830, when a popular uprising in Paris forced Charles X to abdicate.

8. The mechanization of manufacturing depended on all of the following EXCEPT
   a. coal, which provided energy when burned both to run machines and, processed into coke, to smelt iron.
   b. iron, which was a cheap metal which could be formed into durable machines.
   c. steam, which was used to transfer the energy of burning coal into the movements of iron machinery.
   d. horses, which provided the horsepower that was crucial for the operation of heavy machinery.

9. James Watt's steam engines were used to drive all of the following textile machines EXCEPT
   a. Arkwright's water frame.
   b. Crompton's spinning mule.
   c. Savery's atmospheric engine.
   d. Cartwright's power loom.

10. The French Revolution and Napoleonic Wars had all of the following economically harmful effects EXCEPT
    a. destroying resources outright or diverting them into unproductive military activities.
    b. fostering uneconomical French enterprises under the Continental System.
    c. stimulating government intervention in the economy, which left a legacy of over-regulation.
    d. encouraging the overexpansion of British industry, which slumped when peace came.

11. The French Revolution and Napoleonic Wars had all of the following economically helpful effects EXCEPT
    a. removing restrictions on agricultural and industrial production.
    b. establishing uniform commercial regulations and weights and measures.
    c. stimulating government intervention in the economy, resulting in better planning.
    d. improving transportation networks and methods for mobilizing capital.

12. Railroads revolutionized transportation in all of the following ways EXCEPT
    a. by making it possible to move massive quantities of material across country.
    b. by making it possible for people to move across country more freely.
    c. by moving things and people more quickly.
    d. by leading quickly to individual steam-powered vehicles for use on roads.

13. The areas that industrialized after Britain included
    a. Barcelona, Naples, Belgium, northern France, and the Netherlands.
    b. Belgium, northern France, the Netherlands, western Germany, and northern Italy.
    c. Belgium, northern France, the Netherlands, western Germany, and Saxony.
    d. Belgium, northern France, the Netherlands, western Germany, and Naples.

14. Governments contributed to economic development in all of the following ways EXCEPT
    a. developing transportation and communications, including postal services and in many countries railroads.
    b. setting tariffs to protect infant industries from mature foreign competitors.
    c. establishing national banks to facilitate capital mobilization.
    d. directing the allocation of resources according to a rational master plan.

15. Factories contributed to industrialization in all of the following ways EXCEPT
    a. providing a site to house heavy machinery that had to be located in one place.
    b. facilitating the mobilization of capital through stocks and bank loans.
    c. making it possible to closely supervise and discipline the labor force.
    d. acting as centers of gravity for satellite enterprises and residential areas.

16. All of the following are examples of the process of differentiation EXCEPT
    a. the separation of work life from home life because of factories.
    b. the division of economic relationships from social relationships because of money and contracts.
    c. the proliferation of specialized government agencies to oversee different aspects of life.
    d. the growing importance of politics in peoples' thoughts and activities.

17. During the nineteenth century, the idea of the family
    a. became more important, even as the reality of peoples' lives made it more difficult to maintain.
    b. became less important, reflecting the decline of affective ties in industrial society.
    c. did not change much, since the family always had been and remained the foundation of society.
    d. appeared for the first time, since Europeans had earlier considered themselves members of a clan.

18. All of the following were typical roles for nineteenth century European women EXCEPT
    a. small shopkeeper.
    b. factory worker.
    c. professional.
    d. wife and mother.

19. During the early nineteenth century, the standard of living for the great mass of people
    a. clearly rose.
    b. clearly fell.
    c. seems to have remained the same.
    d. changed in different ways for different people

20. During the early nineteenth century, the conditions of life of the great mass of people
    a. clearly got better.
    b. clearly got worse.
    c. seem to have remained the same.
    d. changed dramatically in ways that were generally demoralizing and frequently debilitating.

GUIDE TO DOCUMENTS

I. Metternich Analyzes the Threat to Tranquility

1. All of the following tenets of conservatism are expressed in this document EXCEPT
    a. what the past has bequeathed is worthy of respect.
    b. rulers have a special responsibility to God.
    c. government should stay off peoples' backs.
    d. fathers are the natural heads of families.

2. What, according to Metternich, was so threatening about the middle class?

3. How did Metternich justify conservative principles?

II. Policing Universities—The Carlsbad Decrees

1. According to this document, all of the following threats existed in universities EXCEPT
    a. faculty who expressed subversive opinions.
    b. secret and unauthorized societies.
    c. newspapers and tracts.
    d. armed students.

2. Why might Metternich view universities as threatening?

III. Gladstone Argues for Regulating Railroad Fares

1. What problem did Gladstone foresee if the government did not regulate third-class travel?
    a. People with the cheapest tickets would have to ride in conditions dangerous to their health.
    b. Tickets would be priced too high for poor people to afford to take the train.
    c. Trains would not stop frequently enough for poor people needing search for bread during the journey.
    d. The railroad companies would destroy public faith with their policies toward third-class travel.

2. How does he justify governmental regulation of the railroads?

IV. Reports on the Housing Crisis in France and Germany

1. All of the following characterized rooms described in the reports EXCEPT
    a. they stank.
    b. they were filthy.
    c. they were cold.
    d. they were unlighted.

2. What might bother middle-class observers about the conditions workers lived in?

SIGNIFICANT INDIVIDUALS

1. Louis XVIII
2. Alexander I
3. Prince Metternich (MET-er-nikh)
4. Lord Castlereagh (KAS-el-rā)
5. Prince Talleyrand (TAL-ē-rand)
6. Lajos Kossuth (ku-SOOTH)
7. Thomas Savery
8. Thomas Newcomen
9. James Watt
10. Matthew Boulton

a. Conservative diplomat who dominated peace (1754-1838)
b. Reformist Hungarian leader (1802-1894)
c. British diplomat at Congress of Vienna (1769-1822)
d. Creator of refined steam engines (late 1760s, 1782)
e. Creator of commercial atmospheric engine (1702)
f. Mystical Russian monarch (r.1801-1825)
g. Diplomat for revolution, emperor, and king (1754-1838)
h. Industrialist who focused on steam engines (late 1700s)
i. Developer of piston and cylinder engine (ca.1712)
j. Twice restored King of France (1814-1824)

CHRONOLOGICAL DIAGRAM

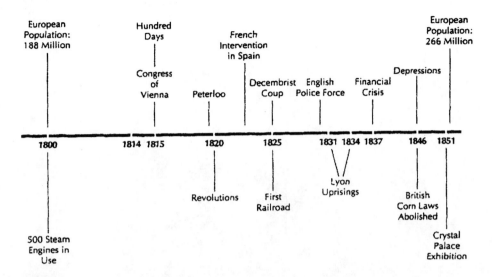

IDENTIFICATION

1. Congress of Vienna
2. Hundred Days
3. Holy Alliance
4. Concert of Europe
5. German Confederation
6. Greek Revolution
7. Carlsbad Decrees
8. *Zollverein* (TSOL-fer-ın)
9. *Carbonari* (kar-bo-NAH-rē)
10. Crystal Palace
11. steam engine
12. factory
13. differentiation
14. standard of living debate

a. International conference that established lasting peace
b. Machine that converts heat into motion
c. Building housing mechanized manufacturing operations
d. Progressive Italian conspirators
e. Prussian-dominated customs union
f. Great Powers' agreement to support conservative regimes
g. Napoleon's last bid for power
h. Spread of specialization among groups and institutions
i. Disagreement among historians about industry's impact
j. Russian monarch's Christian coalition
k. Replacement for Holy Roman Empire
l. Mid-century industrial exhibition
m. German measures to keep students and faculty in line
n. Revolt against Ottomans popular with Romantics

MAP EXERCISES

1. Label the main political divisions of Europe in 1815.
2. Indicate areas of intervention by major European powers during the 1820s.

1. Identify the zones of industrialization and some of the major growing cities of Europe around mid-century.

PROBLEMS FOR ANALYSIS

I. The Politics of Order

1. Evaluate the Congress of Vienna. How realistic were its goals? How well did it meet the problems facing Europe at that time?
2. What methods were used by governments during the restoration to secure both domestic and international order? How effective were they?

II. The Progress of Industrialization

1. Compare patterns of industrialization in Great Britain and on the continent.
2. What role did inventors and machines play in industrialization? What role did governments play in this process?

III. The Social Effects

1. Compare the changing environment, social position, and economic circumstances of the aristocracy, peasantry, working classes, and middle classes.
2. Describe the ideal middle-class family and woman. How does this ideal compare with the realities of the working-class family?
3. Taking into account psychological, social, and economic considerations, did workers benefit from industrialization prior to 1840? Explain.

SPECULATIONS

1. As a conservative political leader in the period 1815–1848, what would be the best policies to follow to maintain social order? Why?
2. Suppose a country were in a position to choose whether to industrialize or not and came to you for advice. Everything considered, what would you recommend? Why?
3. With the advantages of hindsight, what might have been done to ease the pains associated with industrialization and the social change that accompanied it?

TRANSITIONS

In "The Age of Napoleon," the period dominated by the later stages of the French Revolution and Napoleon was examined.

In "Foundations of the Nineteenth Century: Politics and Social Change," the political efforts to maintain domestic and international order are examined. The leaders of the restoration sought to restore the social and international equilibrium of Europe. This meant some compromise between reviving the past and accepting some of the changes of the revolutionary and Napoleonic periods. Politics was chosen as the means for doing this, and politics became the principal concern of the period. During this time much of Europe was becoming transformed by industrialization, accompanied by a variety of social effects. All this would require greater activity and new policies from governments.

In "Learning to Live with Change," the theories of change and movements for reform that arose during the first half of the nineteenth century will be analyzed.

ANSWERS

Self Test

1a; 2d; 3c; 4a; 5c; 6b; 7c; 8d; 9c; 10c; 11c; 12d; 13b; 14d; 15b; 16d; 17a; 18c; 19d; 20d

Guide to Documents

I-1c; II-1d; III-1a; IV-1d

Significant Individuals

1j; 2f; 3a; 4c; 5g; 6b; 7e; 8i; 9d; 10h

Identification

1a; 2g; 3j; 4f; 5k; 6n; 7m; 8e; 9d; 10l; 11b; 12c; 13h; 14i

TWENTY-THREE
LEARNING TO LIVE WITH CHANGE

CHAPTER HIGHLIGHTS

1. Ideologies such as conservatism, liberalism, and early socialism proliferated, offering ways of understanding and dealing with change.
2. A variety of social changes—including class development, population growth, and urbanization—accompanied and interacted with industrialization. Despite the dictates of economic liberalism, governments felt compelled to intervene in social affairs.
3. Reforms in Britain and the revolutions of 1830 mark the beginning of a new age of moderate liberalism in Western Europe.

CHAPTER OUTLINE

I. Ideas of Change
1. Romanticism
2. Social Thought
3. The Early Socialists

II. The Structure of Society
1. Social Classes
2. The Changing Population
3. Social Welfare

III. The Spread of Liberal Government
1. Great Britain
2. The Revolutions of 1830

SELF TEST

1. The romantic movement involved all of the following EXCEPT
   a. a conviction that emotion and experience are the sources of the most profound truths.
   b. a fascination with the power and mystery of nature.
   c. a celebration of the genius that enabled certain people to convey profound insights through art.
   d. a reaffirmation of the central importance of the scientific revolution and Enlightenment.

2. Romanticism was important for all of the following reasons EXCEPT
   a. it nurtured a tremendous outpouring of creative activity in the arts.
   b. it contributed to all three of the modern political ideologies: conservatism, liberalism, and socialism.
   c. it shattered the hold of religion and mysticism on the European mind.
   d. it counterbalanced the rationality and discipline of the scientific world view in modern culture.

3. Edmund Burke's conservatism rejected rationally conceived reform efforts because it held that
   a. "natural" historical development is more reliable than "artificial" plans.
   b. government regulation is the cause, not the cure, of most social problems.
   c. the social order is divinely ordained and cannot change.
   d. privilege is the just reward for superior performance.

4. The classical political liberalism of the early nineteenth century valued which of the following most?
   a. Special interests, like education and the poor.
   b. Maximizing businessmen's profits.
   c. Individual liberty.
   d. Social justice.

5. The classical economic liberalism of the early nineteenth century rested on the assumption that
   a. the government should actively regulate the economy to promote stability and social justice.
   b. individuals pursuing their own interests in a free marketplace will optimize economic activity.
   c. left to its own devices, the "invisible hand" governing the marketplace will bring prosperity to everyone.
   d. liberty in the marketplace included the liberty of workers to form unions to bargain collectively.

6. Riccardo's "iron law of wages" extended the sphere of inexorable economic laws to
   a. social relations.
   b. international economics.
   c. government spending.
   d. metalworking technology.

7. Utilitarianism changed the theoretical underpinnings of liberalism by
   a. substituting maximization of pleasure and minimization of pain for natural rights as its basic justification.
   b. establishing individual liberty as the ultimate purpose of government and source of all good in humanity.
   c. substituting maximization of pleasure and minimization of pain for tradition as the basis for policies.
   d. establishing individual liberty is in essence the maximization of pleasure and the minimization of pain.

8. Utilitarianism, and John Stuart Mills' further thought, had what effect on the liberalism?
   a. It began to advocate reforms for social justice to achieve the greatest good even at the cost of some liberty.
   b. It began to advocate liberty for wider social groups in order to approach the ideal of complete freedom.
   c. It ceased to consider the welfare of the common people as it became the justification for big business.
   d. It began to advocate reforms for social justice as the essential means of achieving individual liberty.

9. Saint-Simon, Fourier, and Owen had all of the following in common EXCEPT
   a. they wanted to create communities in which all people could live well.
   b. they were as opposed to liberalism as they were to the old order.
   c. they were never able to put their ideas into practice.
   d. they were optimistic about human nature.

10. The aristocracy in the nineteenth century was
    a. everywhere in steep decline as industrialization undercut the material bases of its power.
    b. still dominant in industrialized western Europe and was wealthy and influential in the South and East.
    c. still dominant in agrarian southern and eastern Europe and was wealthy and influential in the West.
    d. paradoxically able to bolster its position as industrialization advanced through its political power.

11. The peasantry was substantially affected by all of the following in the early nineteenth century EXCEPT
    a. the commercialization of agriculture.
    b. the mechanization of agriculture.
    c. the decline of the putting-out system
    d. the removal of feudal obligations.

12. Factory workers, including children, typically worked
    a. 8 hours a day, 5 days a week.
    b. 10 hours a day, 5 days a week.
    c. 12-14 hours a day, 6 days a week.
    d. 14-17 hours a day, 6 days a week.

13. The middle class included all of the following groups EXCEPT
    a. artisans and substantial farmers.
    b. shopkeepers, office clerks, and schoolteachers.
    c. merchants, managers, upper bureaucrats, and professionals.
    d. bankers and great industrialists.

14. Europe's population rose from 185 million to 295 million (1800-1870) for all these reasons EXCEPT
    a. increased opportunities for child labor.
    b. a decline in disease-carrying germs.
    c. an increase in the food supply.
    d. a lowering of the age of marriage.

15. The rapid growth of huge cities in the early nineteenth century brought all of the following problems EXCEPT
    a. inadequate sanitation.
    b. abysmal housing.
    c. traffic congestion.
    d. rampant crime.

16. The most effective social welfare measures came from
    a. private charities.
    b. self-help organizations.
    c. the government.
    d. business.

17. While English landlords demanded rents and exported grain from Ireland, the potato famine killed
    a. thousands of Irish peasants.
    b. tens of thousands of Irish peasants.
    c. hundreds of thousands of Irish peasants.
    d. OVER ONE MILLION Irish peasants.

18. English reforms in the 1830s included all of the following EXCEPT
    a. The Reform Bill of 1832 broadening the suffrage and more equitably apportioning representation.
    b. The abolition of slavery in its colonies, the Factory Act limiting child labor, and the Poor Law.
    c. The repeal of the Corn Laws and the victory of the Chartists.
    d. A law granting all resident taxpayers the vote in municipal elections.

19. Of the Chartist call for full democracy and the campaign against the Corn Laws (tariffs keeping food costly)
    a. both succeeded by 1850.
    b. the Chartists succeeded but the Corn Laws remained.
    c. the Corn Laws were repealed, but the Chartists failed.
    d. neither passed because of a growing backlash against reform.

20. The Revolution of 1830 in France had all of the following effects EXCEPT
    a. toppling Charles X and bringing Louis Philippe to the throne.
    b. sparking minor revolts in Italy, Spain, Portugal, and Germany.
    c. precipitating a revolution in Poland that freed it from Russia.
    d. inspiring Belgium to break free from the Netherlands.

21. In the years after 1830, all of the following were true EXCEPT
    a. Louis Philippe ruled as a moderate constitutional monarch.
    b. Mounting tensions in Switzerland led to a civil war that ended with a democratic federation.
    c. Spain became a constitutional monarchy during a dispute of succession to the crown.
    d. England degenerated into anarchy as the government resisted all efforts at reform.

GUIDE TO DOCUMENTS

I. Wordsworth on the Role of the Poet

1. Wordsworth reveals himself to be a romantic in all of the following ways EXCEPT
    a. his preference for rural life.
    b. his appreciation for the value of science.
    c. his view of poetry as the overflow of powerful feelings.
    d. his assertion that poetry of value is produced by a man of unusual natural sensitivity.

2. How does Wordsworth distinguish poetry from science?

II. De Maistre's Opposition to Reform

1. De Maistre supports his argument in all of the following ways EXCEPT
    a. he quotes the philosopher Origen.
    b. he says that every good mind has an instinctive aversion to innovations.
    c. he asserts that the experience of every age validates this aversion.
    d. he shows how the English experience has given them occasion to repent.

2. What sorts of policies would these ideas of de Maistre support?

III. Mill Opposes the Subjection of Women

1. According to Mill, what means were used to subject women?
    a. Fear.
    b. Education.
    c. Seduction.
    d. Coercion.

2. Why, according to Mill, should the subjection of women be ended?

IV. Owen Tells Congress About the Science of Socialism

1. All of the following trends of the early nineteenth century are reflected in this document EXCEPT
    a. the attempt to ground public policy in science.
    b. the conviction that the human mind starts as a "blank slate."
    c. the belief in a universal religion from the heart rather than from dogma.
    d. the reverence for the organic wholeness of premodern civilizations.

2. Why might these ideas be appealing?

V. The Great Charter

1. What is the principal demand of the Chartists presented here?

    a. Universal male suffrage.

    b. Equality of incomes.

    c. Freedom of worship.

    d. Renunciation of the debt.

2. To whom would these demands most appeal, and why? Who might oppose these demands, and why?

SIGNIFICANT INDIVIDUALS

1. Friedrich Schlegel (SHLĀ-gel)
2. Samuel Taylor Coleridge
3. Madame de Staël (stahl)
4. Victor Hugo
5. Hans Christian Andersen
6. J.M.W. Turner
7. Franz Schubert (SHOO bert)
8. Lord Byron
9. Edmund Burke
10. Jeremy Bentham
11. David Ricardo
12. John Stuart Mill
13. Thomas Malthus
14. Count de Saint-Simon (san sē-MON)
15. François Fourier (foo-RYĀ)
16. Robert Owen
17. Louis Philippe (loo-ē fi-LĒP)
18. François Guizot (ghē-ZŌ)
19. Sir Robert Peel
20. Richard Cobden

a. Liberal who sought liberty and social justice (1806-1873)
b. Danish folklorist (1805-1875)
c. Leading German romantic thinker (1767-1845)
d. Socialist who believed in rational planning (1760-1825)
e. Romantic poet critical of church and state (1788-1824)
f. Dominant politician in France in the 1840s (1787-1784)
g. Founder of utilitarianism (1748-1832)
h. France's "July Monarch" (r.1830-1848)
i. French essayist on German philosophy (1766-1817)
j. Socialist who designed ideal community (1772-1837)
k. Manufacturer who worked for free trade (1804-1865)
l. English romantic painter (1775-1851)
m. Leading French romantic novelist (1802-1885)
n. Economist who said population grows faster than food supply (1766-1834)
o. Conservative critic of French Revolution (1729-1797)
p. Propounder of the "iron law of wages" (1772-1823)
q. Romantic composer (1797-1828)
r. English poet of guilt and redemption (1772-1834)
s. Prime minister who gave in on Corn Laws (1788-1850)
t. Industrialist turned socialist (1771-1858)

CHRONOLOGICAL DIAGRAM

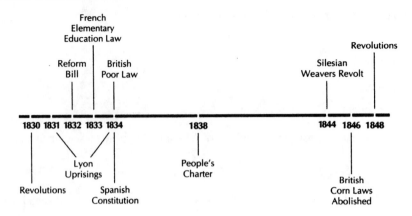

IDENTIFICATION

1. conservatism
2. *livret*
3. iron law of wages
4. cooperatives
5. Poor Law
6. liberalism
7. utilitarianism
8. utopian socialism
9. working class
10. middle class
11. phalanstery
12. Corn Laws
13. Reform Bill of 1832
14. Chartism

a. Ideology with its primary focus on individual liberty
b. People between the aristocracy and the working class
c. "The greatest good for the greatest number"
d. Law extending suffrage and regularizing representation
e. Ideology of resistance to change
f. Theory that competition keeps wages at subsistence level
g. Tariffs on grain that helped landlords, cost consumers
h. English political reform movement
i. English welfare reform of the 1830s
j. Work papers with comments of previous employers
k. Organizations of the poor to pool their resources
l. Proposals for ideal new societies
m. Fourier's ideal society
n. People laboring in factories

MAP EXERCISE

1. Indicate the areas that experienced revolutions in 1830.
2. Indicate those areas adopting liberal institutions in the 1830s and 1840s.

PROBLEMS FOR ANALYSIS

I. Ideas of Change

1. What were the main aspects of early nineteenth-century conservatism?
2. How did liberals evaluate change? What economic policies did they recommend? What role did utilitarianism play in nineteenth-century reform efforts? In what ways did the liberal ideas of John Stuart Mill evolve?
3. What common themes characterize the ideas of Saint-Simon, Fourier, and Owen? Since many of their ideas proved impractical, why are they considered important?

II. The Structure of Society

1. Compare the changing environment, social position, and economic circumstances of the aristocracy, peasantry, working classes, and middle classes.
2. Describe and explain the growth of population during the nineteenth century. How were attitudes toward population changes reflected in the ideas of Thomas Malthus?
3. Considering governmental social policies during the first half of the nineteenth century, how strictly did governments follow the dictates of economic liberalism?

III. The Spread of Liberal Government

1. Compare political liberalism with conservatism.
2. In what ways did the revolutions of 1830 signify the ascendancy of political liberalism in Europe?
3. What policies reflected the growth of liberalism in Britain?

SPECULATIONS

1. Suppose a conservative and a liberal of the early nineteenth century held a debate over the merits of the French Revolution. What would each argue?
2. How might it be argued that even during the 1830s and 1840s, liberalism enjoyed at best only moderate successes?
3. Imagine yourself as a peasant, an industrial worker, a cobbler, and a bank manager. Write a brief description of your living conditions and a typical workday.

TRANSITIONS

In "Foundations of the Nineteenth Century: Politics and Social Change," the conservative policies of governments between 1815 and 1830 as well as the progress of industrialization and the social changes that resulted were examined.

In "Learning to Live with Change," Europeans develop ideologies such as conservatism, liberalism, and early socialism in an effort to cope with rapid change. The structure of European society, particularly where affected by industrialization, is changing. The year 1830 marks the beginning of the spread of liberal institutions, particularly in Western Europe.

In "National States and National Cultures," the revolutions of 1848, the growth of nationalism, and the trends in culture will be examined.

ANSWERS

Self Test

1d; 2c; 3a; 4c; 5b; 6a; 7a; 8a; 9c; 10c; 11b; 12d; 13a; 14a; 15c; 16c; 17d; 18c; 19c; 20c; 21d

Guide to Documents

I-1b; II-1d; III-1b; IV-1d; V-1a

Significant Individuals

1c; 2r; 3i; 4m; 5b; 6l; 7q; 8e; 9o; 10g; 11p; 12a; 13n; 14d; 15j; 16t; 17h; 18f; 19s; 20k

Identification

1e; 2j; 3f; 4k; 5i; 6a; 7c; 8l; 9n; 10b; 11m; 12g; 13d; 14h

TWENTY-FOUR
NATIONAL STATES AND NATIONAL CULTURES

CHAPTER HIGHLIGHTS

1. In 1848 a series of revolutions spread from Paris throughout Europe. Social, political, and national divisions widened as the revolutionaries acquired power, so by August 1849 conservative forces had defeated these inexperienced and divided revolutionary regimes.
2. Nationalism became a powerful and widespread movement.
3. In France, Louis Napoleon combined universal male suffrage, elements of representative government, economic growth, and social benefits with stability and preservation of the social order.
4. Italian and German nationalism—led with political skill by Cavour and Bismarck, respectively—resulted in the unification of Italy in 1861 and of Germany in 1871.
5. Many European states attempted to strengthen and modernize their institutions.
6. Cultural institutions, professions, and styles proliferated during the nineteenth century.

CHAPTER OUTLINE

I. The Revolutions of 1848
1. The Opening Phase
2. The Fatal Dissensions
3. The Final Phase

II. The Politics of Nationalism
1. The Elements of Nationalism
2. A New Regime: The Second Empire in France
3. Nationalism and International Relations
4. A New Nation: The Unification of Italy
5. A New Nation: The Unification of Germany
6. Reshaping the Older Empires

III. Nineteenth-Century Culture
1. The Organization of Culture
2. The Content of Culture

SELF TEST

1. The initial revolutions in 1848 accomplished all of the following EXCEPT
   a. replacing the July Monarchy with the Second Republic in France.
   b. ending Metternich's rule and bringing autonomy for Hungary and promises of a constitution in Austria.
   c. forcing elections to a constituent assembly in Prussia and creation of a German national assembly.
   d. establishing constitutionalism and ending Austrian rule in Italy.

2. The "fatal dissensions" that divided the revolutionary forces included all of the following EXCEPT
   a. conflicts between the middle-class and workers.
   b. national divisions.
   c. the rivalry between Prussia and Austria in Germany.
   d. conflicts between artisans, peasants, and nobles.

3. The "forces of order" used military power to accomplish all of the following EXCEPT
   a. suppressing the workers in Paris during the "June Days."
   b. reimposing serfdom on the peasants of eastern Prussia and Austria.
   c. re-establishing Austrian power and suppressing constitutional regimes in Italy.
   d. crushing uprisings in Austria, Hungary, the Rhineland, Saxony, and Bavaria.

4. The revolutions of 1848 failed for all of the following reasons EXCEPT
   a. the victorious classes and nationalities split after their initial successes because of their divergent interests.
   b. the revolutionaries failed to seize control of the armies and other institutions of coercive power.
   c. the revolutionaries' use of terror alienated the great majority of people.
   d. liberal nations were unwilling to go to war on behalf of revolutionary principles.

5. The revolutions of 1848 accomplished all of the following EXCEPT
   a. creation of an enduring republic in France.
   b. establishment of constitutionalism in Prussia and Piedmont.
   c. emancipation of the serfs in eastern Prussia and Austria.
   d. ending the illusions of broad solidarity among revolutionaries of all classes.

6. Nationalism
   a. is an age-old sentiment arising spontaneously.
   b. is a modern phenomenon often requiring generations of propaganda.
   c. is based on biological differences in appearance and mental ability.
   d. invariably strengthens the power of the state.

7. Nationalism was initially associated with liberalism because both
   a. opposed Europe's dynastic states and cosmopolitan aristocracies.
   b. desired to establish open markets and competitive economies.
   c. wished to strengthen national economies through government programs.
   d. sought the liberation of the individual through limitations on the power of the state.

8. Napoleon III's reign was characterized by all of the following EXCEPT
   a. ambitious programs for social welfare and economic growth.
   b. support for workers' organizations and the right to strike.
   c. gradual liberalization of political institutions.
   d. foreign successes that enhanced France's security.

9. Russia's major opponents in the Crimean War included all of the following EXCEPT
   a. the Ottoman Empire.
   b. Austria.
   c. France.
   d. England.

10. Piedmont joined the allies in the Crimean War in order to
    a. gain an international forum for discussing Italian issues.
    b. participate in a glorious military campaign.
    c. reduce the power of the Ottoman Turks.
    d. cripple Russian naval power in the Mediterranean Sea.

11. The unification of Italy was facilitated by all of the following EXCEPT
    a. Giuseppe Mazzini's nationalist agitation, which raised hopes and expectations despite defeat in 1848.
    b. Cavour's aggressive diplomacy, which gained French support for the defeat of Austria in North Italy.
    c. Garabaldi's "Expedition of the Thousand," which beat the Kingdom of Naples in Sicily and South Italy.
    d. Victor Emmanuel's campaign in 1860, which overran the Papal States and made Rome the capital of Italy.

12. Which of the following was NOT a problem Italy faced after unification in 1861?
    a. Lack of control of Venetia and Rome (which was rectified in 1866 and 1871).
    b. Regional animosities that split the North and the South (still a factor today).
    c. Widespread poverty and corruption (both still problems, particularly the latter).
    d. Lack of ambition in foreign policy (an ongoing aversion to Great Power politics).

13. Prussia's growing dominance of Germany stemmed from
    a. its leaders' historic role as Holy Roman Emperors.
    b. its economic leadership through the *Zollverein*.
    c. Austria's non-German preoccupations.
    d. the dynamism provided by Otto von Bismarck.

14. Bismarck used all of the following wars to unify Germany under Prussia EXCEPT
    a. the 1864 war of Prussia and Austria against Denmark, which established Prussia's equality with Austria.
    b. the Austro-Prussian War of 1866, which enabled Prussia to form the North German Confederation.
    c. the Prusso-Dutch War in 1868, which brought central Germany into confederation with Prussia.
    d. the Franco-Prussian War of 1871, which drew the south German states into the new German Empire.

15. Bismarck's success in unifying Germany while defying Prussia's parliament suggests that
    a. nationalism was stronger than liberalism.
    b. nationalism and liberalism went hand-in-hand.
    c. nationalism ultimately served the purposes of liberalism.
    d. nationalism is ultimately incompatible with liberalism.

16. Russia undertook all of the following reforms in response to the Crimean War EXCEPT
    a. abolishing serfdom.
    b. strengthening village communes.
    c. creating a national parliament.
    d. strengthening provincial representative institutions.

17. The Habsburgs reformed their state after their defeat by the Prussians in 1866 by
    a. creating the long-deferred national parliament.
    b. giving the Hungarians autonomy and equal status.
    c. abolishing serfdom.
    d. breaking the power of the wealthy landlords and merchants.

18. The structure of cultural life changed in the early nineteenth century in all of the following ways EXCEPT
    a. the professionalization of artists.
    b. the merging of high and popular culture.
    c. the proliferation of public facilities for the arts.
    d. the identification of artistic taste with social status.

19. By the middle of the nineteenth century, Romanticism began to give way to
    a. Realism.
    b. Modernism.
    c. Existentialism.
    d. Impressionism.

20. The three elements of Hegel's dialectical schema of history included all of the following EXCEPT
    a. the thesis, the dominant themes and institutions of a civilization.
    b. the antithesis, the reaction against the thesis because no institutions can address all human needs equally.
    c. the parathesis, the partial measures taken by both the thesis and antithesis to accommodate the other.
    d. the synthesis, the reconciliation of the thesis and antithesis, which becomes a new thesis.

GUIDE TO DOCUMENTS

I. The Frankfurt Constitution

1. This document reflects European-wide liberal concerns in all the following ways EXCEPT
    a. the insistence that every citizen has the right to live in any land within the nation.
    b. the abolition of legal class distinctions.
    c. the requirement for due process.
    d. the right of free speech.

2. What circumstances account for the aspects of the document that address specifically German issues?

II. Mazzini's Nationalism

1. How does Mazzini justify the struggle for national unity?
    a. Nationalism enables a person to embrace the whole of humanity, which is in the end one great nation.
    b. Nations are the associations that will fulfill God's purpose by enabling people to work together for good.
    c. Nations are grounded in geography, which, as an aspect of nature, is inherently good.
    d. Nations are the fulfillment of the statebuilding process begun by the royal dynasties of early modern times.

2. In what ways does Mazzini attempt to broaden and strengthen the appeal of his ideas?

III. Bismarck's Social Program

1. Bismarck gives all of the following reasons for introducing social welfare legislation EXCEPT
    a. to undercut support for socialism.
    b. to fulfill the duty of a patriarchal state.
    c. to promote the Christian religion in Germany.
    d. to keep business from being saddled with the burden.

2. In what ways is his approach different from that of liberals?

SIGNIFICANT INDIVIDUALS

1. Napoleon III
2. Lord Palmerston
3. Franz Joseph I
4. Giuseppe Mazzini (joo-SEP-ē maht-TSĒ-nē)
5. Count Camillo Cavour (ka-VOOR)
6. Giuseppe Garibaldi (ga-ri-BALL-dē)
7. Otto von Bismarck (fon)
8. Franz Liszt (list)
9. Giuseppe Verdi (Vair-dē)
10. Richard Wagner (VAHG-ner)
11. Charles Dickens
12. Honoré de Balzac
13. Sir Walter Scott
14. George Eliot (Mary Cross)
15. Gustave Flaubert (flō-BAIR)
16. George Sand
17. Jules Michelet (mēsh-LE)
18. George Hegel
19. Gustave Corbet (koor-BE)

a. The "Iron Chancellor" who unified Germany (1815-1898)
b. Romantic adventurer who helped unify Italy (1807-1882)
c. Assertive British foreign minister (1784-1865)
d. Author of acid account of a young middle-class wife's aimless existence (1821-1880)
e. French historian of France's dramatic fight for freedom (1798-1874)
f. English author of romantic novels about chivalry (1771-1832)
g. Early Italian nationalist (1805-72)
h. Realist painter (1819-1877)
i. Not quite his uncle (1808-1873)
j. Influential female English novelist (1819-1880)
k. Statesman who led Italian unification (1810-1861)
l. Emperor of the "Dual Monarchy" (r.1848-1916)
m. Leading Italian composer of operas (1813-1901)
n. Leading German composer of operas (1813-1883)
o. Author focusing on "the human comedy" (1799-1885)
p. English novelist who exposed social ills (1812-1870)
q. Philosopher of dialectical view of history (1770-1831)
r. Virtuoso pianist and romantic composer (1811-1886)
s. Influential female French novelist (1803-1876)

CHRONOLOGICAL DIAGRAM

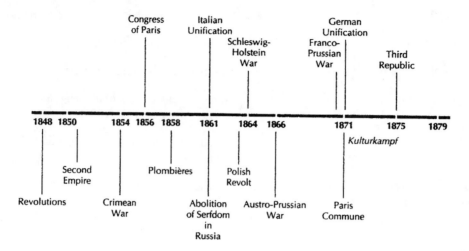

IDENTIFICATION

1. June Days
2. nationalism
3. Congress of Paris
4. Don Pacifico affair
5. Italian National Society
6. Expedition of the Thousand
7. "blood and iron"
8. North German Confederation
9. Battle of Sedan
10. *mir* (mēr)
11. Dual Monarchy
12. *Kulturkampf*
13. Bibliothèque Nationale (bib-lē-ō-TĀK nas-yo-NAL)
14. Realism

a. What Bismarck said would decide the issues of the day
b. Artistic style focusing on ordinary people and situations
c. International conference that ended the Crimean War
d. Swashbuckling campaign to liberate Sicily from Naples
e. Emotional tie between individual and large ethnic group
f. Bismarck's struggle against the Catholic Church
g. Russian peasant commune
h. The union of Austria and Hungary under Habsburg crown
i. Civil war between French workers and bourgeoisie
j. British defense of Jewish citizen in Greece
k. Fight in which Napoleon III was captured
l. Nationalist group that supported Piedmont
m. Prussian dominated union created after the Austro-Prussian War
n. The French national library

MAP EXERCISES

1. Identify those areas experiencing revolts and revolutions in 1848.

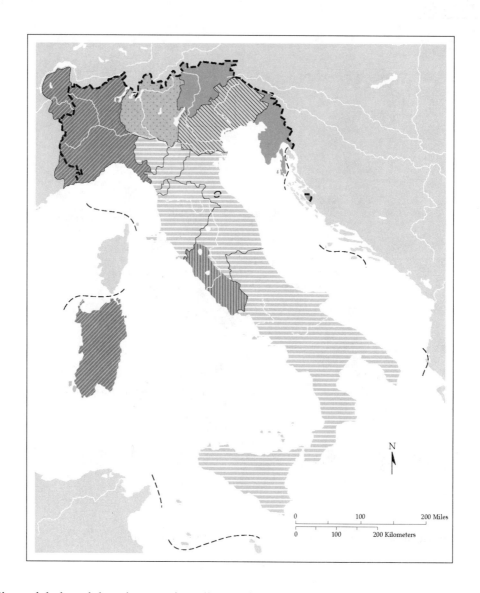

1. Indicate, label, and date the steps in Italian unification between 1859 and 1870.

1. Indicate and label the steps in German unification between 1815 and 1871.

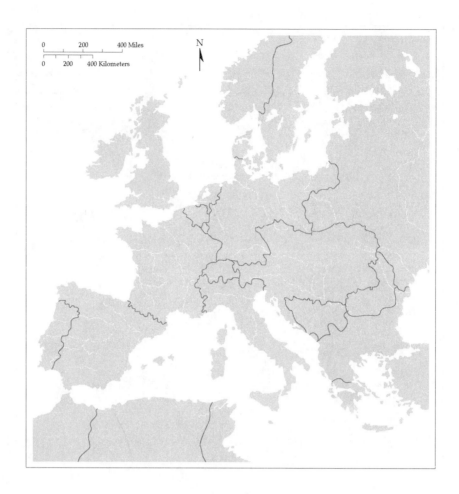

1. Label the main political divisions of Europe in 1871.
2. In what ways has this political map of Europe changed since 1848?

PROBLEMS FOR ANALYSIS

I. The Revolutions of 1848

1. Explain the origins of the revolutions of 1848.
2. Why were the forces of order able to crush the revolutionary regimes in 1848 and 1849?

II. The Politics of Nationalism

1. Why did nationalism develop into such a powerful force during the nineteenth century? In what ways did both governments and revolutionary groups use nationalism?
2. Considering the manner in which he rose to power, his domestic policies, and his foreign policies, evaluate the political style of Napoleon III. How does the liberal trend during the second half of his reign fit into this?
3. What forces and events led to the Crimean War? In what ways was this war significant?
4. Compare the roles of Cavour and Bismarck in the unification of their respective countries.
5. Explain the ways in which the Russian government attempted to modernize during the reign of Alexander II.

III. Nineteenth-Century Culture

1. In what ways did nineteenth-century culture differ from the culture of the eighteenth and seventeenth centuries?
2. Compare Romanticism and Realism.
3. Analyze the relationship of the Enlightenment and the Romantic movement in terms of Hegel's dialectic of thesis, antithesis, and synthesis.

SPECULATIONS

1. How would German or French liberals who had supported the revolutionary cause in 1848 have felt about developments between 1849 and 1871? Explain.
2. Do you think the end—German unification—justified the means used by Bismarck? Explain.
3. Nineteenth-century thought, art, literature, and music remains extremely popular today. What was so extraordinary about this culture?

TRANSITIONS

In "Learning to Live with Change," intellectual and social trends as well as the spread of liberal politics during the first half of the nineteenth century were analyzed.

In "National States and National Cultures," we see demands for reform rise in the revolutions of 1848, only to succumb to the forces of order. Nationalism spreads everywhere, most powerfully revealed in the unifications of Italy and Germany and in the demands to strengthen national states. Meanwhile cultural styles and institutions proliferate throughout Europe as the arts become more national and public.

In "European Dynamism and the Nineteenth-Century World," stress will be on new economic growth, scientific developments, and modern imperialism.

ANSWERS

Self Test

1d; 2c; 3b; 4c; 5a; 6b; 7a; 8d; 9b; 10a; 11d; 12d; 13b; 14c; 15a; 16c; 17b; 18b; 19a; 20c

Guide to Documents

I-1a; II-1b; III-1c

Significant Individuals

1i; 2c; 3l; 4g; 5k; 6b; 7a; 8r; 9m; 10n; 11p; 12o; 13f; 14j; 15d; 16s; 17e; 18q; 19h

Identification

1i; 2e; 3c; 4j; 5l; 6d; 7a; 8m; 9k; 10g; 11h; 12f; 13n; 14b

TWENTY-FIVE
EUROPEAN DYNAMISM AND THE NINETEENTH-CENTURY WORLD

CHAPTER HIGHLIGHTS

1. Economic growth and industrial progress accelerated during the second half of the nineteenth century. The period was also one of great demographic growth, particularly in cities.
2. Science, ideas of progress, and knowledge of the non-European world proliferated during this period.
3. Particularly after the 1860s, European nations—motivated by a combination of economic, cultural, and political considerations—engaged in a vast imperial expansion.

CHAPTER OUTLINE

I. The Economics of Growth
   1. The Second Industrial Revolution
   2. The Demographic Transition

II. The Knowledge of Nature and Society
   1. The Conquests of Science
   2. Social Science and Ideas of Progress

III. Europe and the World
   1. The Apparent Decline of Colonial Empires
   2. Europe's Growing Engagement Overseas

IV. Imperialism and European Society
   1. The Meanings of Imperialism
   2. Explanations of Imperialism
   3. Imperialism's and European Society
   4. Conquests of the New Imperialism

SELF TEST

1. The technological foundations of the second industrial revolution included all of the following EXCEPT
   a. new smelting processes that made steel the most commonly used metal.
   b. the proliferation of chemicals used in products ranging from dyes to fertilizers.
   c. the rapidly spreading use of electricity in lighting and to power machinery.
   d. the use of coal to run increasingly powerful steam-engines made from iron.

2. Germany surpassed England as Europe's leading industrial power for all of the following reasons EXCEPT
   a. Germany industrialized later, so its factories were larger and more up-to-date.
   b. in the late nineteenth century England's economy began to contract after decades of growth.
   c. the German educational system produced the administrators and engineers vital to modern industry.
   d. German cartels rationalized production and supported aggressive marketing around the world.

3. The proportion of people in agriculture could decline because all of the following raised productivity EXCEPT
   a. fertilizers.
   b. machinery.
   c. imports.
   d. specialization.

4. Europe's population grew from 295,000,000 people in 1870 to 450,000,000 in 1914 because
   a. The birth rate rose while the death rate stayed steady.
   b. The death rate fell while the birth rate stayed steady.
   c. The birth rate fell, but the death rate fell far more.
   d. The death rate rose but the birth rate rose far more.

5. By 1900, the proportion of people making their living from farming in England was
   a. 43 percent.
   b. 35 percent.
   c. 22 percent.
   d. 8 percent.

6. Fundamental scientific advances were made in all of the following fields EXCEPT
   a. thermodynamics.
   b. electromagnetism.
   c. astronomy.
   d. chemistry.

7. The nineteenth century's interest in historical development was expressed in all of the following EXCEPT
   a. Comte's positivism, which said civilization progresses through successive stages of intellectual ability.
   b. Marx's position that changes in the means of production cause class struggles that drive social evolution.
   c. Darwin's theory, which said species evolve through competition to adapt to a changing environment.
   d. Spencer's philosophy that progress from complexity to simplicity is a universal necessity.

8. Marxism proved strongly influential for all of the following reasons EXCEPT
   a. it was logically compelling and relevant to all of the social sciences
   b. it gave scientific validity to the rejection of industrial society.
   c. it provided a hard-headed moral criticism of its opponents.
   d. it claimed the prestige of science and gave the security of determinism.

9. Social Darwinism's application of Darwin's theory of natural selection to society was most often used to justify

    a. the class struggle.

    b. social inequality.

    c. monopolistic practices.

    d. moral reform efforts.

10. Latin America seemed to show that Europe's age of empire was past for all of the following reasons EXCEPT

    a. from 1804 to 1824 France and Spain lost virtually all of their colonies.

    b. during subsequent conflicts, France and England made no moves to take over.

    c. Mohammed Ali's revolt against the Ottomans led to modernizing internal reforms

    d. independence did not destroy Europe's cultural influence or profitable trade.

11. Europe's presence around the world was increased informally by all of the following EXCEPT

    a. explorers and missionaries.

    b. bureaucrats and planners.

    c. merchants and bankers.

    d. local officials and emigrants.

12. In the late nineteenth century, imperialism referred to

    a. the legacy of the Roman Empire.

    b. Europe's mid-century informal influence overseas.

    c. Europe's direct rule over much of the non-European world.

    d. economic and cultural domination with or without political rule.

13. Imperialism was connected to the development of capitalism by

    a. the division of the world into a wealthy and powerful economic core and a poor and powerless periphery.

    b. the growth of a world market from which all continents and peoples profited equally.

    c. the advances which Europe first learned from non-Europeans, but then used against them.

    d. the spread of free trade as the concept of the free market was adopted throughout the world.

14. Early theorists of imperialism focused on

    a. militarism.

    b. economic interests.

    c. nationalism.

    d. sociological factors.

15. More recent theories about imperialism have focused on all of the following EXCEPT

    a. the machinations of great financiers.

    b. the growth of the popular press.

    c. explorers and missionaries.

    d. ambitious soldiers and local officials.

16. Economic and technological developments that fostered imperialism include all of the following EXCEPT

    a. the growing demand for a wide range of raw materials.

    b. the need for telegraph posts and naval coaling stations.

    c. modern weapons and tropical medicines.

    d. the radio and the airplane.

17. Historians today study imperialism for the light it sheds on all of the following issues in European culture EXCEPT

    a. regionalism.

    b. class.

    c. race.

    d. gender.

18. All of the following social groups in Europe particularly encouraged imperialism EXCEPT

    a. the aristocracy, which wanted subject peoples to lord over.

    b. religious societies, which wanted to convert heathen souls.

    c. peasants, who saw colonies as more land available to farm.

    d. merchants, who wanted to increase their business activities.

19. Imperialism was popular among Europeans for all of the following reasons EXCEPT

    a. it offered opportunities for heroic initiative and noble self-sacrifice.

    b. it gave even lower-class Europeans people they could feel superior to.

    c. it served as a national unifier, aligning conservative groups, businessmen, and the common people.

    d. it encouraged social reforms at home in order to mobilize public support for expansion abroad.

20. Europeans established direct rule or substantial control in all of the following regions EXCEPT

    a. Africa.

    b. Southeast Asia.

    c. Latin America.

    d. China.

21. Japan was able to avoid conquest by the Europeans by

    a. isolating itself from the outside world.

    b. modernizing its economy and government.

    c. buying off the imperial powers with access to raw materials.

    d. playing the different imperial powers off against each other.

GUIDE TO DOCUMENTS

I. Making the Deals That Created a Cartel

1. According to this document, how did a cartel work?
   a. Each firm bid on a project, and the contract went to the lowest bidder.
   b. The government allocated portions of a contract among all firms willing to match the lowest bidder.
   c. The firms agreed to collaborate in offering high bids in exchange for an agreed proportion of the work.
   d. The government kept customs duties high in order to give domestic firms the highest prices possible.

2. What sorts of economic pressures faced managers during this period of industrialization?

II. Huxley's Social Darwinism

1. How do Huxley's views relate to Darwin's theories?
   a. Huxley shows man's current condition and nature to be an outgrowth of natural selection.
   b. Huxley relates man's intelligent energy to the cosmic force that pervades the universe.
   c. Huxley demonstrates that natural selection has led to the development of ethical man.
   d. Huxley suggests that the same characteristics that helped man in nature will help develop society.

2. What attitudes and social policies might logically flow from these views presented by Huxley?

III. The Interpretation of Imperialism

1. According to Hobson, what is the source of imperialism?
   a. The inevitable laws of history.
   b. The greed of a powerful minority.
   c. The interests of the nation as a whole.
   d. The workings of genuine democracy.

2. What problems does Hobson say imperialism caused?

3. What, according to Schumpeter, is the main source of imperialism?
   a. The fusing of nationalism and militarism.
   b. The "inner logic" of capitalist development from the methods of early capitalism.
   c. The alliance of precapitalist classes oriented to war with the military interests among the bourgeoisie.
   d. The fear of the precapitalist classes that in the long run the climate of the modern world will destroy them.

4. What does Schumpeter suggest will be the ultimate fate of the imperialists, and hence imperialism?

5. How does Mommsen indicate recent research has changed our understanding of imperialism?

   a. Focus on informal imperialism, the imperialists' mind-set, and the role of non-European developments.

   b. A renewed emphasis on the endogenous policies and economic structures of the industrial states.

   c. A questioning of the role of the periphery and sociological or socio-economic explanations.

   d. Creation of a nostalgia for the days when the European powers were responsible for developments.

6. According to Mommsen, what should a modern understanding of imperialism include?

SIGNIFICANT INDIVIDUALS

1. Nicolas Carnot (kahr-NŌT)
2. Michael Faraday
3. James Maxwell
4. Dmitri Mendeleev (men-de-LĀ-ef)
5. Louis Pasteur (pas-TUR)
6. Auguste Comte (kont)
7. Karl Marx
8. Friedrich Engels
9. Charles Darwin
10. Herbert Spencer
11. William Gladstone
12. Cecil Rhodes

a. Biologist who originated theory of evolution (1809-1882)
b. Marx's sidekick (1820-1895)
c. Physicist of electricity, magnetism, and light (1831-1879)
d. Social scientist and radical agitator who created "scientific socialism" (1818-1883)
e. Theoretician of thermodynamics (1796-1832)
f. Positivist philosopher (1798-1857)
g. Imperialists' imperialist in Southern Africa (1853-1902)
h. Philosopher of progress (1820-1903)
i. Scientist who studied magnetism (1791-1867)
j. Germ killing biologist (1822-1895)
k. Chemist who developed periodic table (1834-1907)
l. Reluctant imperialist (1809-1898)

CHRONOLOGICAL DIAGRAM

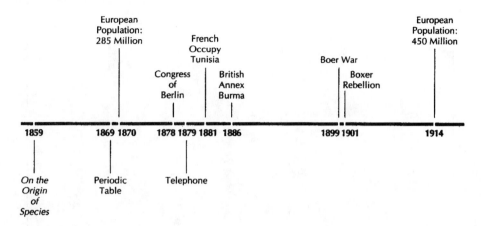

IDENTIFICATION

1. the second industrial revolution
2. the demographic transition
3. Marxism
4. positivism
5. Darwinism
6. social Darwinism
7. imperialism
8. Boxer Rebellion
9. Boer War

a. Period of economic boom when growth became the norm
b. Theory of evolution through natural selection
c. Simultaneous decline of birth and death rates
d. Tendency of Europeans to take over the rest of world
e. Application of theory of evolution to human society
f. Theory of social progress based on dialectical materialism
g. Philosophy of historical development based on knowledge
h. Revolt of white South Africans against British
i. Revolt of Chinese against foreigners

MAP EXERCISES

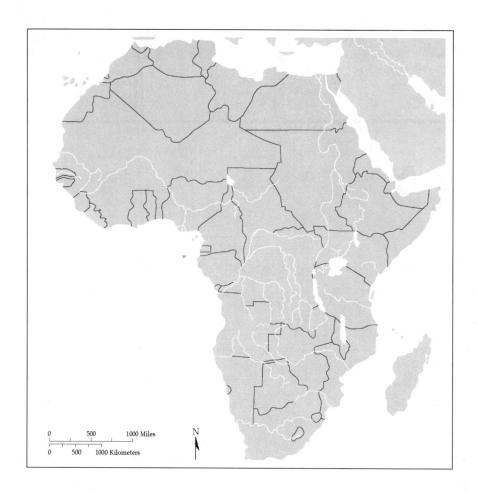

1. Indicate the possessions acquired by each of the European powers by 1914.

1. Indicate the possessions and spheres of influence acquired by the various powers in 1914.

PROBLEMS FOR ANALYSIS

I. The Economics of Growth

1. Explain the comparatively spectacular industrial expansion of Germany and the relative loss of British industrial supremacy between 1870 and 1914.

2. How do you explain the sharp rise in European population between 1870 and 1914 despite falling birthrates in Western European countries and massive migration of people overseas?

II. The Knowledge of Nature and Society

1. How did scientific discoveries and development of theories of progress relate to each other?

II. Europe and the World

1. In what ways was European presence around the globe expanding? How do you explain this?

III. Modern Imperialism

1. Discuss alternative explanations for late nineteenth-century imperialism. Develop what you feel is the best explanation.

2. Was colonial conquest part of a conscious, long-range scheme by various European governments, or was it simply something they fell into by accident or through the force of events? Explain.

SPECULATIONS

1. Suppose there were a debate between Karl Marx and Herbert Spencer. What would be the principal areas of agreement and disagreement?

2. In what ways might it be argued that the closing decades of the nineteenth century represented the peak of European power and accomplishment? What disturbing trends were present during these decades?

3. Suppose you were an imperialist such as Cecil Rhodes. How would you justify imperialism to an African? How might an African reply?

TRANSITIONS

In "National States and National Cultures," the growth of nationalism and the unification and strengthening of national states during the three decades following the revolutions of 1848 were traced, and the culture of the nineteenth century was analyzed.

In "European Dynamism and the Nineteenth-Century World," three main developments during the half-century from the 1860s to 1914 are examined. The period was marked by tremendous economic growth and increasing population. Scientific and intellectual developments multiplied, increasing Europe's knowledge of the world. Finally, the causes, growth, and effects of imperialism are examined.

In "The Age of Progress," cultural developments, attack on liberal civilizations, and domestic politics in the decades before World War I will be explored.

ANSWERS

Self Test

1d; 2b; 3c; 4c; 5d; 6c; 7d; 8b; 9b; 10c; 11b; 12c; 13a; 14b; 15a; 16d; 17a; 18c; 19d; 20c; 21b

Guide to Documents

I-1c; II-1a; III-1b; III-3c; III-5a

Significant Individuals

1e; 2i; 3c; 4k; 5j; 6f; 7d; 8b; 9a; 10h; 11l; 12g

Identification

1a; 2c; 3f; 4g; 5b; 6e; 7d; 8i; 9h

TWENTY-SIX
THE AGE OF PROGRESS

CHAPTER HIGHLIGHTS

1. The decades between 1860 and 1914 witnessed a great growth of popular culture, social institutions, and cultural styles.
2. Marxists, radicals, Christians, and conservatives attacked the ideals and practices of liberalism.
3. Almost all European nations experienced difficulties adjusting to social change that were reflected in political conflict and efforts for reform.

CHAPTER OUTLINE

I. The Belle Époque
1. Popular Culture
2. "The Woman Question"
3. The Arts

II. Attacks on Liberal Civilization
1. Working Class Movements
2. The Christian Critique
3. Beyond Reason

III. Domestic Politics
1. Common Problems
2. France: The Third Republic
3. Germany: The Reich
4. Italy: The Liberal Monarchy
5. Russia: Defeat and Revolution
6. Austria-Hungary: The Delicate Balance
7. Spain: Instability and Loss of Empire
8. Great Britain: Edging Toward Democracy

SELF TEST

1. All of the following developments in popular culture occurred in the late nineteenth century EXCEPT
    a. the professionalization of entertainment.
    b. the institutionalization of leisure.
    c. the regionalization of amusements.
    d. the attainment of almost universal literacy.

2. Women's' organizations included all of the following types EXCEPT
    a. moderate groups, which worked to amelioration of the conditions of women's lives.
    b. women's trade unions, which fought for better pay and working conditions.
    c. radical groups, which sought fundamental legal and social equality.
    d. feminarchist groups, which sought to reverse the sex roles and subordinate men.

3. By 1910, women in major industrial countries had gained all of the following EXCEPT
    a. increased rights to property.
    b. a share in decisions affecting their children.
    c. a right to participate in public affairs.
    d. the right to vote.

4. During this period, the dominant trend in the arts was from meticulous reproduction of external reality to
    a. criticism of contemporary society and its mores.
    b. depiction of the artists' subjective perceptions.
    c. expression of the artist's interior emotions.
    d. experimentation with abstract form and color.

5. Marx's relationship with other socialists can best be characterized as
    a. warm collegiality.
    b. doctrinaire combativeness.
    c. cool indifference.
    d. fawning obsequiousness.

6. As socialist parties became important in Europe, they gradually developed
    a. more moderate rhetoric combined with more moderate policies.
    b. more moderate rhetoric to mask their radical policies.
    c. a combination of radical rhetoric and moderate policies.
    d. a combination of radical rhetoric and radical policies.

7. Churches attacked liberalism on all of the following grounds EXCEPT
    a. it put too much emphasis on the individual.
    b. it was indifferent to moral issues.
    c. it failed to uphold the rights of property.
    d. it put too much emphasis on material things.

8. The intellectual attack on liberal civilization was led by all of the following EXCEPT
    a. Sorel, who rejected bourgeois rationalism in favor of violence and the will.
    b. Bergson, who rejected reason in favor of feelings, spontaneity, and common endeavor.
    c. Nietzsche, who attacked its Christian "slave morality" and emphasized will and uninhibited supermen.
    d. Thiers, who argued that liberal civilization would have to adopt the violence of its foes to defend itself.

9. Anti-Semitism increased during this period for all of the following reasons EXCEPT
    a. Jews were identified in the popular mind with the dislocations and inequities of modern society.
    b. a secret Jewish congress had met to organize a conspiracy to take over control the world.
    c. crude adaptations of Darwinism gave a veneer of scientific respectability to ancient antipathies.
    d. conspiracy theories gave concrete and simple explanations for the baffling pace of social change.

10. Political conservatism was supported by all of the following social groups EXCEPT
    a. the peasantry, which was threatened both by rising equipment costs and foreign competition.
    b. the aristocracy, which saw its traditional power eroded by the rise of businessmen and professionals.
    c. the workers, who wanted to return to the security and prosperity of the early nineteenth century.
    d. shopkeepers and independent artisans, who faced the competition of large retailers and factories.

11. All of the following were examples of the increasing scale of social organizations in this period EXCEPT
    a. the expansion and bureaucratization of the civil service.
    b. the growth and organization of the Catholic Church.
    c. the appearance of big businesses and the organization of cartels.
    d. the growth of labor unions and professional societies.

12. The turmoil following Napoleon III's defeat was resolved by
    a. the suppression of the Paris Commune.
    b. the founding of the Third Republic.
    c. the restoration of Charles X's Bourbon monarchy.
    d. the restoration of Louis Philippe's Orleans dynasty.

13. The Dryfus affair was important because
    a. the conservatives lost prestige when it was shown that they were vilifying an innocent man.
    b. the liberals lost prestige when it was shown that they were defending a guilty man.
    c. the conservatives lost prestige when people realized that they were acting out of anti-Semitism.
    d. the liberals lost prestige when people realize that they were defending a Jew.

14. The fundamental political flaw in Imperial Germany was that
    a. it had a conservative monarchy, which grew frustrated that power was held by the liberal parliament.
    b. it had an increasingly liberal parliament, but ultimate power was held by the conservative monarchy.
    c. it had never been resolved where ultimate power lay: in the hands of the parliament or the monarchy.
    d. it was so dominated by conservatives that it refused to make any concessions to businessmen or workers.

15. Germany's conservatives used political leagues to campaign for all of the following EXCEPT
    a. high tariffs.
    b. an overseas empire.
    c. the creation of a world-class navy.
    d. the extension of social welfare programs.

16. The socialist's strength in Germany was shown by all of the following EXCEPT
    a. the Social Democratic party became the largest in 1912.
    b. the labor unions had 2.5 million members in 1912.
    c. the Social Democrats were secure enough to adopt Bernstein's moderate policy.
    d. the Social Democrats sustained an extensive workers' subculture.

17. The basic flaw in Italian politics was that the government had difficulty pursuing popular programs because it
    a. was elected by the wealthy and had to cut deals with special interests to keep its parliamentary majority.
    b. was committed to modernizing the nation while balancing the budget.
    c. found it necessary to become increasingly restrictive as time went on.
    d. continually faced disasters in foreign policy that distracted it from domestic affairs.

18. Reforms were forced on Nicholas II of Russia by all of the following EXCEPT
    a. the ignominious defeat by the Japanese in the Russo-Japanese War of 1904.
    b. "Bloody Sunday," the incident which touched off the Revolution of 1905.
    c. a general strike that brought the nation's economy to a standstill.
    d. the "October Manifesto" which granted the nation a constitution.

19. Russian politics after the Revolution of 1905 were characterized by all of the following EXCEPT
    a. a parliament, the Duma, which gave the nation its first taste of representative government.
    b. repeated disbanding of the Duma and tinkering with the electoral laws to produce a conservative majority.
    c. concessions to the middle class and workers that moved Russia quickly into the European mainstream.
    d. reforms initiated by the government of education, administration, local government, and the economy.

20. Austria-Hungary's politics during this period can best be described as
    a. stalemated.
    b. progressive.
    c. regressive.
    d. vigorous.

21. Liberal and Conservative competition for votes in Britain resulted in all of the following reforms EXCEPT
    a. revised codes for housing and urban sanitation and the removal of restrictions on unions.
    b. disestablishment of the Anglican Church and restriction of abuses by landlords in Ireland.
    c. creation of a national education system and army reforms outlawing the purchase of commissions.
    d. the extension of the vote to all adult males through the Reform Bills of 1867 and 1885.

22. All of the following brought or threatened violence in Britain EXCEPT
    a. strikes by dissatisfied workers demanding better pay and working conditions.
    b. the House of Lords' veto of Lloyd George's "people's budget."
    c. demonstrations by suffragettes demanding the vote for women.
    d. the mobilization of Irish Protestants after Parliament granted Ireland home rule.

GUIDE TO DOCUMENTS

I. G. B. Shaw Explains the Appeal of Popular Theater

1. The document reveals all of the following about late nineteenth century cultural life in England EXCEPT

    a. it was divided by cost between the "high" culture of the rich and the popular culture of the masses.

    b. it was an immensely profitable business.

    c. popular culture was inferior in artistic quality to high culture.

    d. the common people reacted more spontaneously and respectfully to the artists than the rich did.

2. What was the function of the drama critic? For whom was Shaw writing this review? What was he trying to convey to his readers?

II. Bakunin on Why He Opposes the State

1. According to Bakunin, what is the difference between (1) private morality and (2) the state's morality?

    a. (1) is based on respect for human freedom and dignity; (2) is based on serving the state's power.

    b. (1) is based on the private relationship of the individual to God; (2) is based on serving the state's power.

    c. (1) is based on respect for human freedom and dignity; (2) is based on the state's divine mission.

    d. (1) is vitiated by religious dogmas; (2) is based on the status of the state as a divine institution.

2. What arguments does Bakunin make against the State? Why might these arguments be appealing?

III. The Argument of Anti-Semitism

1. The document makes all the following charges against the Jews EXCEPT

    a. while Christians do most of the work, Jews get most of the profits.

    b. Jews control the press and thereby control what the Germans think.

    c. Jews subvert Germanic ideals with cosmopolitan culture.

    d. Jews eat Christian babies.

2. To whom might these views be appealing? Why?

IV. Emmeline Parkhurst on Women's Rights

1. The ultimate justification for women's suffrage made in this document is that women need the vote in order to

    a. establish more equitable divorce laws.

    b. make men live up to their responsibilities to women.

    c. represent their own interests and take responsibility for their actions.

    d. contribute their intelligence to policy-making so that the world can be made a better place.

2. How does Parkhurst connect the emancipation of women with the larger goals of improving the human race?

SIGNIFICANT INDIVIDUALS

1. Emile Zola
2. Claude Monet (mō-NE)
3. Vincent van Gogh (van gō)
4. Maria Montessori (mon-te-SOR-ē)
5. Mikhail Bakunin
6. Georges Sorel
7. Pius IX
8. Henri Bergson
9. Friedrich Nietzsche (NĒ-che)
10. Eduard Bernstein
11. Adolphe Thiers (tyair)
12. William II
13. Nicholas II
14. Peter Stolypin (sto-LĒ-pēn)
15. William Gladstone
16. Benjamin Disraeli (diz-RĀ-lē)

a. Liberal British prime minister (1809-1898)
b. Conservative British prime minister (1804-1881)
c. Italian educator (1870-1952)
d. Philosopher who decried Christian "slave morality" and called for amoral supermen to lead humanity (1844-1900)
e. Founder of the Third Republic (1797-1877)
f. Painter of dazzling, often mystical works (1853-1890)
g. Tsar of Russia during 1905 Revolution (r.1894-1917)
h. French syndicalist (1847-1922)
i. Master of the naturalist novel (1840-1902)
j. Pope who decried the errors of our times (r.1846-1878)
k. French philosopher of spontaneity (1859-1941)
l. Brash young German emperor (r.1888-1918)
m. Russian anarchist (1814-1876)
n. Masterful impressionist painter (1840-1926)
o. Prime minister after the 1905 Revolution (r.1906-1911)
p. German "revisionist" socialist (1850-1932)

CHRONOLOGICAL DIAGRAM

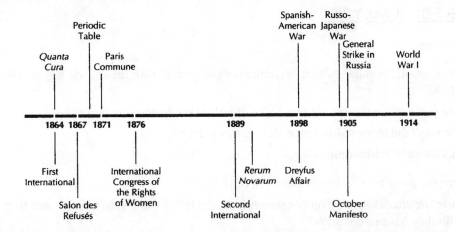

IDENTIFICATION

1. International Congress of the Rights of Women
2. Naturalism
3. Salon des Refusés (re-fyoo-sā)
4. Symbolism
5. First International
6. anarchism
7. anti-Semitism
8. *Quanta Cura*
9. Paris Commune
10. Third Republic
11. Dreyfus case
12. Second Reich (rīkh)
13. *Reichstag* (RĪKHS-takh)
14. Social Revolutionaries
15. 1905 Revolution
16. October Manifesto
17. *Duma* (DOO mah)

a. French government formed after the collapse of Napoleon III's Second Empire
b. Papal catalogue of the "principal errors of our time"
c. Irrational anger toward and scapegoating of Jews
d. Style of poetry linking surface and spiritual realities
e. Controversy over Jewish officer charged with treason
f. Russian radicals combining populism and terrorism
g. The German Empire
h. Exhibition of paintings barred from official exhibit
i. Revolutionary uprising in Russia after the war with Japan
j. First multinational socialist gathering dominated by Marx
k. Group that met at the Paris exposition of 1878
l. Artistic movement depicting life in objective detail
m. Russian parliament
n. Tsar's constitutional concessions after general strike
o. Revolutionary uprising in French capital
p. Philosophy rejecting the imposed authority of the state
q. German parliament

PROBLEMS FOR ANALYSIS

I. The Belle Époque

1. In what ways did popular culture become much more national, urban, and organized during this period?
2. How do you explain the proliferation of cultural styles during this period?
3. In what ways did modern life began during this period?

II. Attacks on Liberal Civilization

1. Analyze the growth of Marxism and its appeal.
2. Compare the attacks leveled on liberalism from the left, from conservatives, and from others. What do they have in common?

III. Domestic Politics

1. What common threads run through political developments in several nations during this period?
2. What forces resulted in the shift to more liberal institutions and policies in Russia during the first decade of the twentieth century? Compared to Western Europe, how liberal were these reforms?

SPECULATIONS

1. Considering political, social, and cultural developments of the times, in what ways might it be argued that the closing decades of the nineteenth century were one of the best times to live?
2. In a debate between late nineteenth-century liberals and their opponents, what points might be most powerfully made on each side?

TRANSITIONS

In "European Dynamism and the Nineteenth-Century World," the second industrial revolution, the new developments in science, and the expanding presence and influence of Europeans around the globe were examined.

In "The Age of Progress," cultural and political developments during the final decades of the nineteenth century and the first decade of the twentieth century are examined. Culturally this was the *Belle Époque,* marked by growing urban leisure activities and a proliferation of artistic styles. At the same time movements to reject liberalism rose from all sides. Political systems struggled to maintain social control and reform; relative if fragile peace prevailed.

In "World War and Democracy," World War I and the immediate postwar period will be analyzed.

ANSWERS

Self Test

1c; 2d; 3d; 4b; 5b; 6c; 7c; 8d; 9b; 10c; 11b; 12b; 13a; 14b; 15d; 16c; 17a; 18d; 19c; 20a; 21d; 22b

Guide to Documents

I-1c; II-1a; III-1d; IV-1d

Significant Individuals

1i; 2n; 3f; 4c; 5m; 6h; 7j; 8k; 9d; 10p; 11e; 12l; 13g; 14o; 15a; 16b

Identification

1k; 2l; 3h; 4d; 5j; 6p; 7c; 8b; 9o; 10a; 11e; 12g; 13q; 14f; 15i; 16n; 17m

SECTION SUMMARY
THE NINETEENTH CENTURY 1789–1914
CHAPTERS 20–26

CHRONOLOGICAL DIAGRAM

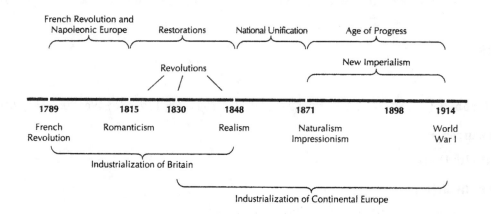

MAP EXERCISES

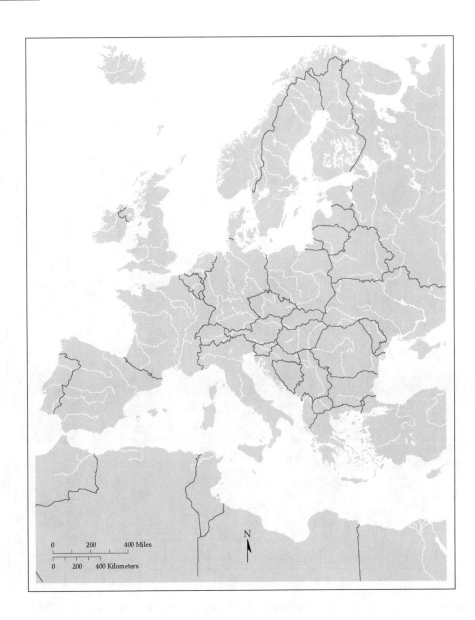

1. Indicate the spread of industrialization and approximate dates during the period 1789–1914.

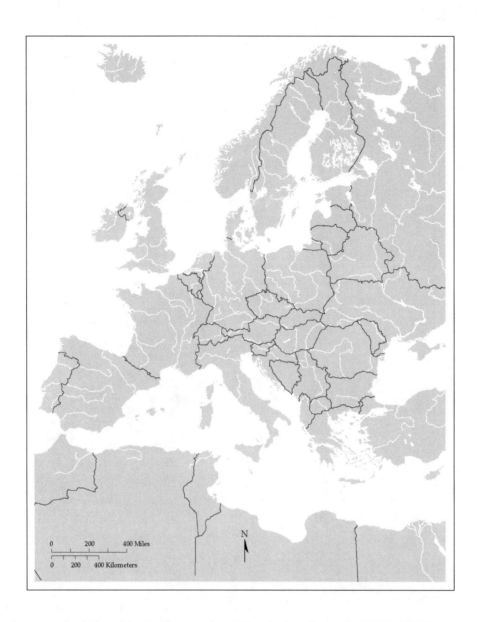

1. Indicate the spread of liberal institutions and policies during the period 1789–1914.
2. Indicate the spread of national unification movements and the establishment of new national states during the period 1789–1914.
3. Indicate areas experiencing revolutions, and the dates involved, during the period 1789–1914.

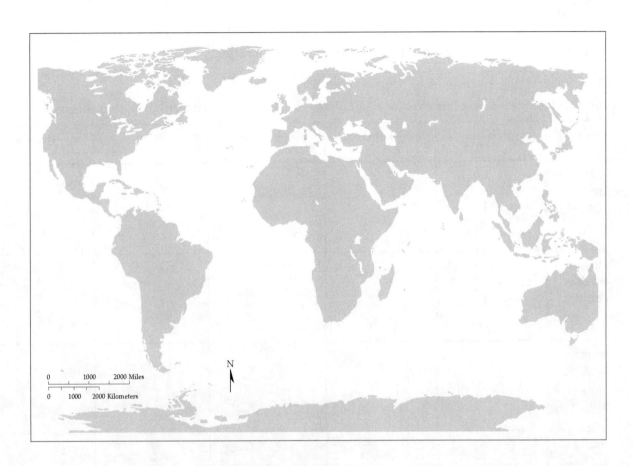

1. Indicate those areas of the world under the political control of Europe and the United States in 1914.

BOX CHARTS

Reproduce the Box Chart in a larger format in your notebook or on a separate sheet of paper. For a fuller explanation of the themes and how best to find material, see Introduction.

Chart 1:

It is suggested that you devote one page for each column (i.e., chart all seven themes for each country on a page).

Country Themes	Britain	France	Germany	Italy	Russia
Social Structure: Groups in Society					
Politics: Events and Structures					
Economics: Production and Distribution					
Family Gender Roles Daily Life					
War: Relationship to larger society					
Religion: Beliefs, Communities, Conflicts					
Cultural Expression: Formal and Popular					

Chart 2:

It is suggested that you devote one page for each row (i.e., chart each of the seven themes on a separate page; you may want to use blank or graph paper and turn the page sideways). Note major national differences, but keep the focus on development in Europe as a whole.

Periods Themes	1789	1815	1848	1875	1914
Social Structure: Groups in Society					
Politics: Events and Structures					
Economics: Production and Distribution					
Family Gender Roles Daily Life					
War: Relationship to larger society					
Religion: Beliefs, Communities, Conflicts					
Cultural Expression: Formal and Popular					

CULTURAL STYLES

1. Look at the pictures on pages 711, 712, 713, 717, 720, 722, and 723? What do they show about the evolution of the revolution? How does their composition reflect the events that they portray?

2. Examine the portraits of Napoleon on pages 746, 736, and 753. How do they reflect the different stages in his rise to power? To what extent do they seem to be a conscious attempt to contribute to that process?

3. Contrast the portraits in question 2 and the picture on page 748 with the pictures on pages 751, 757, 758, and 763. What do they convey about the changing perceptions of Napoleon's reign? How do their differing styles contribute to the message they are trying to convey?

4. Contrast the picture on page 771 with the one on page 804. What different attitudes and values do they express? How do their composition and use of color contribute to the message they convey?

5. Look at the pictures on pages 791, 823, 887, and 971. What do they show about the changing scale of industry from the first to the second industrial revolution?

6. Compare the paintings on pages 847, 850, and 866 with the photographs on pages 837, 946, and 953. What do the paintings convey that the photos don't? What do the photos convey that the paintings don't? Thinking of modern news photographs, to what extent were the limitations of the photos in comparison to the paintings technical problems having to do with the cumbersomeness of the early equipment, and to what extent are they inherent in the differences between the media?

7. Look at the picture on page 844. What do the setting and details convey about the nature of German unification?

8. Compare the pictures on page 688, 689, 804, 808, 879, 926, 934, and 936 (both). What do they show about the transition from Classicism to Romanticism to Realism to Subjectivism? Thinking back to the art of the Renaissance, Baroque, and Classical periods, and the changes from one to the other, in what ways were changes in the nineteenth century more fundamental than the earlier stylistic developments?

TWENTY-SEVEN
WORLD WAR AND DEMOCRACY

CHAPTER HIGHLIGHTS

1. Diplomacy became increasingly constrained by a hardening system of alliances, the politics of nationalism and imperialism, and domestic demands, leading to an unwanted war that nevertheless served as a unifying force for societies under strain.
2. By the end of 1914 the struggle turned into a long war of attrition supported by massive extension of governmental control.
3. Leaders of the Paris Peace Conference, facing conflicting interests, produced treaties that supposedly would reflect the optimistic liberal ideals of the Allies, but which nevertheless outraged some and left all dissatisfied.
4. After an early period of some instability, most European nations followed a democratic path during the 1920s.

CHAPTER OUTLINE

I. The Coming of World War
1. Bismarck's System of Alliances
2. The Shifting Balance
3. The Outbreak of World War
4. The Origins of World War

II. The Course of the War
1. The Surprises of the First Two Years
2. Adjustment to Total War
3. The Great Trials of 1917–1918
4. The Effects of World War I

III. The Peace
1. The Revolutionary Situation
2. The Peace Treaties

IV. Postwar Democracy
1. The New Governments
2. The Established Democracies
3. International Relations

SELF TEST

1. The "Bismarckian system" of the 1870s and 1880s involved all of the following EXCEPT
    a. secret pacts pledging mutual defense or neutrality in case of attack by another power.
    b. diplomatic maneuvers to keep France isolated to avoid the threat of its revenge over Alsace-Lorraine.
    c. the careful but ruthless use of warfare to gain national ends against diplomatically isolated enemies.
    d. maintaining peace among the European powers by allowing them balanced gains outside Europe.

2. After Bismarck's dismissal, his successors' perpetuated his system but failed to
    a. maintain secret defensive pacts.
    b. keep France isolated.
    c. use the threat of war.
    d. keep a balance in overseas gains.

3. The chief cause of Germany's diplomatic failure was
    a. its abrasive manner and aggressive policies.
    b. its repeated betrayal of its allies.
    c. the cunning of its enemies and perfidy of its allies.
    d. its inability to back its words with power.

4. The reason that England moved into the anti-German camp was it felt threatened by
    a. Germany's colonial gains.
    b. the German navy.
    c. German's huge army.
    d. Germany's economic competition.

5. World War I started when all of the following came to a head EXCEPT
    a. Austria-Hungary's rivalry with Russia in the Balkans.
    b. France's desire to get revenge against Germany for 1870.
    c. Britain's long-standing commitment to Belgian neutrality.
    d. Italy's colonial rivalry with France in North Africa.

6. The war did not go as expected because
    a. armies moved with an unanticipated speed because of advances in technology.
    b. it did not end quickly since technology favored the defense.
    c. naval battles rather than battles on land proved to be the decisive factor.
    d. the German army did not live up to the reputation it had gained in 1870.

7. With the repeated, costly failure of offensives designed to break through enemy lines, the Germans
   a. went on the defensive, hoping to gain victory through the economic collapse of the Allies.
   b. tried to negotiate, offering a peace based on the status quo at the beginning of the war.
   c. tried attrition, trading German lives for Allied lives until, they hoped, the Allies would give up.
   d. appealed to the Americans, hoping they would attack the Allies from behind.

8. In contrast to the stalemate on the Western front, the Eastern front saw sweeping movements of armies that
   a. gradually wore Germany down.
   b. ended up at the same point they began.
   c. forced Austria-Hungary out of the war.
   d. eventually brought the collapse of Russia.

9. Deaths in different campaigns included all of the following EXCEPT
   a. 25,000 Frenchmen and 20,000 Germans in the Battle of the Marne in September 1914.
   b. Over 300,000 Frenchmen and almost 300,000 Germans at Verdun from February to July 1916.
   c. More than 300,000 on each side during the Battle of the Somme from July to November 1916.
   d. More than 1,000,000 Russians in Brusilov's offensive during 1916.

10. The war at sea
    a. played no significant role in the war.
    b. was decided by the clash of the British and German fleets at Jutland.
    c. was important for the economic effects of Britain's blockade and Germany's U-boat campaign.
    d. was important mainly because Britain was able to launch amphibious attacks on the German flanks.

11. The British, French, and Germans mobilized themselves for protracted war by all of the following EXCEPT
    a. taking control of the economy to maximize production for the war effort.
    b. suppressing domestic dissent.
    c. drawing women into traditionally male jobs to replace men conscripted into the army.
    d. abolishing social distinctions in order to promote social cohesion.

12. With the collapse of the Russian autocracy and the entry of the United States, the Allies styled the war as
    a. the West against the East.
    b. democracy against monarchy.
    c. capitalism versus communism.
    d. a struggle against German dictatorship.

13. During 1918 all of the following occurred EXCEPT
    a. the communist government in Russia made a separate peace with Germany.
    b. Germany launched a desperate offensive in the West that brought it substantial gains.
    c. the Western Allies held and, reinforced by fresh American troops, began to throw the Germans back.
    d. the German Emperor realized that the war was lost and concluded a humiliating peace with the Allies.

14. The First World War had all of the following effects EXCEPT
    a. it caused widespread and profound psychological disillusionment, social instability, and political turmoil.
    b. it transformed Europe from creditor to debtor status.
    c. it killed between 10,000,000 and 13,000,000 people, including about 4,000,000 civilians.
    d. it increased the status of the aristocracy and middle class relative to the workers and peasantry.

15. The war destroyed all of the following empires EXCEPT
    a. the Austro-Hungarian.
    b. the Ottoman.
    c. the German.
    d. the British.

16. The Treaty of Versailles imposed all of the following on Germany EXCEPT
    a. territorial losses in the West, the East, and overseas.
    b. heavy reparations.
    c. complete disarmament.
    d. responsibility for the war.

17. Which of the following was NOT a consideration in drawing the new borders in Eastern Europe?
    a. Allegiances during the war.
    b. The will of the people.
    c. The principle of legitimacy.
    d. Security interests.

18. The new German republic faced all of the following problems EXCEPT
    a. the leftist "Spartacist" uprising in 1919.
    b. hyper-inflation.
    c. rightist paramilitary activity.
    d. war with Poland and Hungary.

19. Economic recovery and assimilation of social changes led to stable democracy in all of the following EXCEPT
    a. France and Britain.
    b. Scandinavia and the Low Countries.
    c. Italy, Greece, and Hungary.
    d. Germany and Czechoslovakia.

20. All of the following were effective in reducing the sources of international tensions during the 1920s EXCEPT
    a. the Dawes Plan, which regularized Germany's reparations payments and arranged for American loans.
    b. the Locarno Pact, which settled Germany's border with France.
    c. the Kellogg-Briand pact, which renounced war "as an instrument of national policy."
    d. the Washington Naval Conference, which limited the number of major warships in the worlds' navies.

GUIDE TO DOCUMENTS

I. The Terms of the Triple Alliance

1. The nature of the terms of the treaty presented here can best be described as
    a. defensive.
    b. offensive.
    c. optimistic.
    d. imperialistic.

2. Do you think these arrangements promoted the possibilities for peace or made war more probable? Why?

3. Under the terms stated here, do you think Italy was justified in not supporting Austria-Hungary in 1914?

II. Meet the "Khaki Girls"

1. What effect of the war on English society revealed in this passage?
    a. radicalizing.
    b. leveling.
    c. demoralizing.
    d. trivializing.

2. According to the authors of this account, in what ways do men and women workers differ? In what ways, according to the authors, are women and men workers similar?

III. Wilfred Owen Describes Trench Warfare

1. What does this reveal about the nature of the fighting during World War I?
    a. It was fast-paced.
    b. It was grueling.
    c. It was sporadic.
    d. It was exciting.

2. What might be the lasting effects of someone going through the sorts of experiences described by Owen?

IV. German Inflation

1. German inflation distorted social and economic life in all of the following ways EXCEPT
    a. by making it impossible for Germans to afford first class services in their own country.
    b. by discrediting the values of thrift and discipline that had characterized the German middle class.
    c. by increasing the gap between the wage-earning majority and the rich minority who lived off investments.
    d. by making worthless the pensions of handicapped veterans, robbing them of their promised reward.

2. What does this document reveal about social conditions during this period of post-war Germany?

SIGNIFICANT INDIVIDUALS

1. Archduke Francis Ferdinand
2. David Lloyd George
3. Georges Clemenceau (kluh-man-SO)
4. Vittorio Orlando
5. Joseph Joffre (zhoffr)
6. Henri Philippe Pétain (pā-tin)
7. Paul von Hindenburg
8. Erich von Ludendorff (loo-den-dorf)
9. Fredinand Foch (fush)
10. Woodrow Wilson
11. Friedrich Ebert (Ā-bert)
12. Mustafa Kemal (ke-mahl)
13. Rosa Luxemburg
14. Raymond Poincaré (pwan-ka-RĀ)
15. Aristide Briand (brē-AN)
16. Gustav Stresemann

a. Austrian whose assassination started the war (1863-1914)
b. French commander in chief, 1914 to 1917 (1852-1931)
c. German chief of staff from 1916 to 1918 (1847-1934)
d. American president in war and peace talks (r.1913-1921)
e. Prime minister who led Britain to victory (1863-1945)
f. Socialist first president of Weimar Republic (r.1919-1925
g. Supreme Allied commander in 1918 (1851-1929)
h. French prime minister in 1920s (1860-1934)
i. Italian premier at Versailles peace talks (r.1917-1919
j. German socialist leader killed as Spartacist (1870-1919)
k. Premier who led France to victory (1841-1929)
l. French commander in chief in 1918 (1856-1951)
m. Dominant politician in Weimar Republic (1878-1929)
n. Leading French politician of the 1930s (1862-1932)
o. German master strategist late in war (1865-1937)
p. Turkish leader after fall of Ottoman Empire (1881-1938)

CHRONOLOGICAL DIAGRAMS

Diagram 1

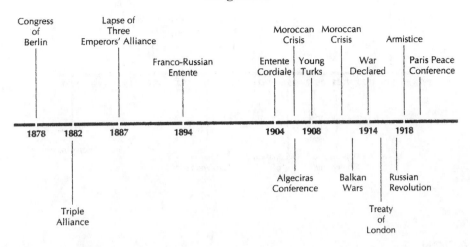

Diagram 2

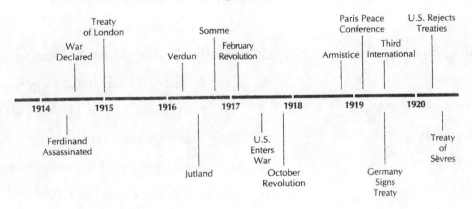

Diagram 3

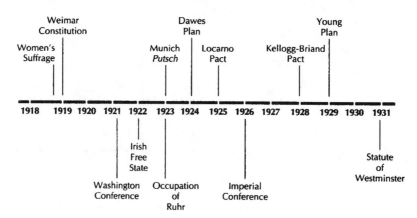

IDENTIFICATION

1. Triple Alliance
2. Triple Entente (an-TANT)
3. system of alliances
4. Schlieffen plan (shlē-fin)
5. war of attrition
6. Treaty of Brest-Litovsk
7. Fourteen Points
8. League of Nations
9. war guilt clause
10. Sinn Fein (shin FAYN)
11. Spartacists (SPAR-ta-sists)
12. Dawes Plan
13. Locarno Pact
14. Imperial Conference of 1926

a. Strategy of wearing the enemy down
b. Agreement that took Russia out of the war
c. Militant Irish nationalist party
d. Treaty between Germany, Austria-Hungary, and Italy
e. Gathering that reorganized British Empire
f. Section of Versailles treaty that blamed war on Germany
g. Revolutionary German socialists
h. German strategy to win the war against France quickly
i. Alliance of France, Russia, and Britain
j. Major agreement to help keep the peace singed in 1925
k. Wilson's plan for peace
l. First attempt at World Government
m. International order based on formal coalitions
n. American brokered adjustment to economics of peace

MAP EXERCISES

1. Indicate the main diplomatic alliances on the eve of World War I in 1914.

1. Indicate the principal territorial changes resulting from the settlements following World War I.

PROBLEMS FOR ANALYSIS

I. The Coming of World War

1. Trace the breakdown of the Bismarckian system and the development of a new system of alliances between 1878 and 1914.
2. Assess the relative importance of diplomacy, imperialism, and nationalism in causing World War I.
3. Historians argue that this was an unwanted war, yet its outbreak was almost universally greeted with joy. How do you explain this contradiction? Does it point to domestic tensions as a major cause for the war, or was this reaction simple patriotism?

II. The Course of the War

1. Characterize the military strategy applied in World War I. How does this strategy compare with the technical effectiveness of the offense and the defense?
2. Why did the Allies win the war?
3. Analyze the nonmilitary effects of the war. How was the role of government changed? What social and psychological changes resulted?

III. The Peace

1. What were the main goals of the Paris Peace Conference? How much agreement was there on these goals?
2. Evaluate the treaties. How well did they meet the problems they were designed to deal with? Was there cause by 1920 to be optimistic about the treaties? Explain.

IV. Postwar Democracy

1. In what ways might the years after World War I be considered a triumph for liberalism?
2. What trends of the postwar period might now, in retrospect, seem problem-filled and ominous?

SPECULATIONS

1. What might diplomats and political leaders have done between 1878 and 1914 to prevent the outbreak of World War I? What realistic options were open to them?
2. How would you rewrite the peace treaties to secure a more just and lasting peace? Do you think your suggestions would have been accepted by the leading negotiators? Explain.
3. What might have been done to alter the course of German developments during the 1920s to diminish the chances for the rise of Nazism in the 1930s?

TRANSITIONS

In "The Age of Progress," the cultural and political developments of the period between 1860 and 1914 were examined, with particular emphasis on intellectual attacks on liberalism.

In "World War and Democracy," the diplomatic events and underlying tensions leading to World War I are studied. On one level, World War I originated from the diplomatic system that grew over the preceding half-century. On a deeper level, domestic pressures tied to the economic and social changes of the same period led to the war. The course and consequences of the war are then examined. Those most adjusted to these changes survived best during and after the war; the Austrian and Russian empires failed. Finally, the peace treaties and developments among the democracies in the postwar period are analyzed. For a while, democracy, international cooperation, and relative prosperity seemed to spread, but so were disturbing trends.

In "The Great Twentieth-Century Crisis," some of these trends—the revolutions, the growth of totalitarianism—are examined as well as the growth of a distinctive twentieth-century culture.

ANSWERS

Self Test

1c; 2b; 3a; 4b; 5d; 6b; 7c; 8d; 9a; 10c; 11d; 12b; 13d; 14d; 15d; 16c; 17c; 18d; 19c; 20c

Guide to Documents

I-1a; II-1b; III-1b; IV-1c

Significant Individuals

1a; 2e; 3k; 4i; 5b; 6l; 7c; 8o; 9g; 10d; 11f; 12p; 13j; 14h; 15n; 16m

Identification

1d; 2i; 3m; 4h; 5a; 6b; 7k; 8l; 9f; 10c; 11g; 12n; 13j; 14e

TWENTY-EIGHT
THE GREAT TWENTIETH-CENTURY CRISIS

CHAPTER HIGHLIGHTS

1. In Russia, revolutions in 1917 brought the Bolsheviks under Lenin to power. After winning a bitter civil war, the Bolsheviks began introducing communism into the new Soviet Union.
2. In Italy, Mussolini led the Fascist party to power and established an increasingly powerful dictatorship.
3. Uncertainty and pessimism characterized dominant trends in thought and culture during the first decades of the twentieth century.
4. Authoritarian regimes, often approaching fascism, spread during the 1920s and 1930s in eastern, central, and southern Europe.
5. A devastating and seemingly uncontrollable Depression debilitated the West during the 1930s.
6. In Germany, Hitler's Nazism, building upon anti Semitism and militaristic struggle, became a terrifying form of totalitarianism.
7. In the Soviet Union totalitarianism under Stalin succeeded—through the five-year plans and the Great Purges—in collectivizing agriculture, industrializing the country, and eliminating all potential enemies.

CHAPTER OUTLINE

I. Two Successful Revolutions

1. Revolution in Russia
2. Toward a Communist Society
3. Italian Fascism

II. The Distinctive Culture of the Twentieth Century

1. Freudian Psychology
2. The Humanities
3. The Sciences
4. Public Culture

III. The Retreat from Democracy

1. Authoritarian Regimes
2. The Great Depression

IV. Totalitarianism: Nazi Germany and the U.S.S.R.

1. Hitler's Germany
2. Stalin's Soviet Union

V. Democracy's Weak Response
1. Divisive Social Change
2. The Argument for Liberty
3. Domestic Politics
4. The Failures of Diplomacy

SELF TEST

1. The Bolsheviks came to power in Russia through all of the following EXCEPT
   a. the tsar lost power to the Provisional Government because of disillusionment with war and poverty.
   b. the Provisional Government lost power to the Bolsheviks because of disillusionment with war and poverty.
   c. the Bolsheviks seized and held on to power because they were phenomenally determined and disciplined.
   d. the mystic Rasputin was angry with the royal family, so he sold his soul for the power to destroy it.

2. The New Economic Policy
   a. imposed communism on Russia through collectivization and the five year plans.
   b. permitted small businesses and peasant enterprise to help rebuild the country.
   c. intensified and extended "war communism" to cope with the threat of civil war.
   d. accepted private ownership of major industries while taking over the small businesses.

3. Fascism manifested discontents rooted in all of the following EXCEPT
   a. war weariness.
   b. inflation.
   c. unemployment.
   d. disillusionment about the peace treaty.

4. All of the following characterized Fascist rule in Italy EXCEPT
   a. the leader of the party, Mussolini, ruled by decree.
   b. opponents were outlawed and suppressed.
   c. economic life was regulated by confederations of employers and workers.
   d. people's lives gradually improved as per capita output and the real wage rose.

5. Freud's theories made people particularly uncomfortable for all of the following reasons EXCEPT
   a. they involved sex, which was a taboo subject in nineteenth century middle class society.
   b. they made explicit and therefore threatened the delicate balance of forces within people's psyches.
   c. they undercut the security people derived from feeling that they could control the world rationally.
   d. they came at a time when all other facets of the culture were validating the liberal consensus.

6. Philosophy, literature, and the arts all shared
   a. a conviction that reason held the key to understanding and organizing human affairs.
   b. a fear that the irrational would subvert all that was valuable in modern Western civilization.
   c. a denial of the ability of human reason to connect people to the ultimate sources of truth.
   d. an insistence that reason and feeling must co-exist and are resolvable eclectically on an ad hoc basis.

7. The Newtonian synthesis was shattered by all of the following insights into physical reality EXCEPT
   a. Einstein's discovery that length, mass, and time are not absolute, but change with relative velocity.
   b. Heisenberg's proof that there are physical limits to what can be known.
   c. both scientists' realization that physical reality is affected by the observer's condition and intervention.
   d. the fact that Newtonian physics is accurate enough for most human purposes.

8. Less spectacular advances in other sciences included all of the following EXCEPT
   a. the isolation of viruses and the development of antibiotics in medicine.
   b. discoveries about the nature of heredity and their application to agriculture.
   c. the application of statistical tools to the study of human society.
   d. the refinement of thermodynamics and the explosive creativity of the Bauhaus.

9. The 1920s saw the arrival of an exciting new art form:
   a. movies.
   b. the novel.
   c. opera.
   d. poetry.

10. European culture was characterized by all of the following EXCEPT
    a. science and the arts alike lost contact with the everyday experience of and relevance to ordinary people.
    b. the creators of high culture began to see popular culture as an invaluable source of inspiration.
    c. movies, magazines, and books proliferated and popularized themes and images drawn from high culture.
    d. some high cultural trends, like the celebration of violence, paralleled similar trends in popular culture.

11. Democratic governments in eastern and southern Europe generally lost power to regimes that were
    a. communist, modeled on the Soviet Union.
    b. fascist, combining authoritarian rule with popular mobilization.
    c. authoritarian, based on a strongman fronting for the old upper classes.
    d. Nazi, following the German lead in implementing fascism plus race war.

12. By 1932 the Great Depression was so bad that all of the following were true EXCEPT
    a. the world was producing only two-third as many manufactured goods as in 1928.
    b. 13,000,000 Americans, 6,000,000 Germans, and 3,000,000 English people were unemployed.
    c. Germany stopped paying reparations, and the ex-Allied countries stopped paying their war debts.
    d. a series of communist revolutions broke out across Europe, inspired by Russia's success.

13. Hitler and the Nazis were able to come to power in Germany because of all of the following EXCEPT
    a. Hitler's messianic speaking style, his angry yet hopeful message, wealthy patrons, and the use of violence.
    b. the long-term dissatisfactions of many Germans with the defeat in World War I and social upheavals after.
    c. the immediate misery caused by the Great Depression, which the Weimar republic seemed unable to solve.
    d. Hitler sold his soul to the Devil to avenge having to work as a male prostitute in pre-World War I Vienna.

14. Nazi rule during the 1930s was characterized by all of the following EXCEPT
    a. creation of a party structure paralleling, and increasingly dominating, state organizations.
    b. outlawing, arresting, and murdering opponents within and outside the party, and especially targeting Jews.
    c. gradual softening of the party's ideological positions as it adapted to the responsibilities of power.
    d. massive government works projects and rearmament, which quickly reduced unemployment.

15. After Lenin died, leadership in the Soviet Union went to
    a. Stalin.
    b. Trotsky.
    c. Bukharin.
    d. a collective leadership including all three.

16. The Five Year Plans accomplished all of the following EXCEPT
    a. collectivizing agriculture, despite massive resistance, deportations, and famines that killed millions.
    b. making the USSR the world's third largest producer, increasing steel by five times and electricity by 24!
    c. increasing literacy rates from below 50 percent to above 80 percent, and raising the standard of living.
    d. ending the need for strict political controls because of widespread satisfaction with the economic growth.

17. The divisive social changes plaguing the democracies included all of the following EXCEPT
    a. increasing capitalization costs for farmers.
    b. increasingly structured and pressured factory production lines.
    c. increasing ethnic tensions caused by mass migrations of laborers.
    d. increasing distance between elite and popular cultures.

18. Individual freedom, human dignity, and social justice were championed by all of the following EXCEPT
    a. Alfred Rosenberg and Giovonni Gentile
    b. Nikolai Berdyaev and Martin Buber.
    c. W. H. Auden, Thomas Mann, and André Malraux.
    d. Karl Barth and Jacques Maritain.

19. Keynesian economics insisted that faced with depression, instead of cutting spending to balance their budgets,
    a. governments should "prime the pump" with deficit spending to stimulate demand for goods and services.
    b. individuals should use credit cards to buy things in order to stimulate demand for goods and services.
    c. businesses should take out loans and expand their operations to capitalize on low prices and wages.
    d. banks should lend aggressively in order to stimulate increased production for the market.

20. The Spanish Civil War was a crucial development during the 1930s because

  a. it demonstrated that the democracies lacked the will to oppose aggression in Europe.

  b. it created a strong new Fascist power that played a vital role in World War II.

  c. it offered an opportunity for the West and Russia to cooperate as they would later in the war.

  d. it proved that aggression would be met by forceful action by the democracies, deterring war.

GUIDE TO DOCUMENTS

I. Two Accounts of Revolution in Russia

1. What accounts for the different implications of the two accounts?

  a. Sorokin was an opponent of the Bolsheviks; Reed was a supporter.

  b. Reed was an eyewitness, while Sorokin was conveying second-hand information.

  c. Reed was in one part of the palace; the people Sorokin was talking to were in another.

  d. The Bolsheviks had reason to protect the property while they also had reasons to kill the people.

2. What sorts of problems facing historians are revealed by these two accounts?

II. Fascist Doctrine

1. Fascism is distinguished here from both liberalism and socialism by all of the following EXCEPT

  a. it denies that human society exists to promote the welfare of the individual.

  b. it denies the importance of the future in comparison to the values embodied in the past.

  c. it denies the fundamental equality of people and the right of all to vote at least indirectly.

  d. it denies the possibility of peace and celebrates the effects of war.

2. In this excerpt, what are the bases for fascism's appeal?

III. The Futurist Manifesto

1. The passage exalts all of the following EXCEPT

  a. energy and speed.

  b. struggle and audacity.

  c. morality and women.

  d. violence and destruction.

2. What specific aspects of the doctrine outlined here account for the name "futurism?"

3. In what ways might Futurism be seen as sympathetic to some of the ideas and spirit of fascism?

IV. Spengler's View of History

1. Spengler states or implies all of the following EXCEPT

    a. democracy is a corrupt system established by capitalists to facilitate their rule.

    b. the power of capitalism can only be defeated through violence.

    c. history will vindicate the stronger because might can only go to those who are right.

    d. a dictatorship that will overthrow the rule of money is already in development.

2. How, like Futurism, might this excerpt be viewed as reflective of the environment and experience of the decades following 1914?

V. Goebbels' Populist View of German Culture

1. According to Goebbels, art suffered from all of the following problems EXCEPT

    a. it was dominated by outmoded concepts and esthetics.

    b. it was practiced in a small, ingrown community.

    c. it was patronized by the bored and snobbish rich.

    d. it was supervised by Jewish intellectuals.

2. To whom and in what way is he appealing? Why might he feel it is important to gain support of these cultural views?

SIGNIFICANT INDIVIDUALS

1. Sigmund Freud (froid)
2. C.G. Jung (yoong)
3. Emile Durkheim
4. Max Weber (vā-ber)
5. Marcel Proust (proost)
6. Franz Kafka
7. James Joyce
8. Oswald Spengler
9. José Ortega y Gasset
10. Bertrand Russell
11. Ludwig Wittgenstein
12. Virginia Wolf
13. Albert Einstein
14. Max Planck
15. Werner Heisenberg (HĪ-zen-bergk)
16. Aleksander Kerensky
17. Nikolai Lenin
18. Leon Trotsky
19. Benito Mussolini (moos-o-LĒ-nē)
20. Antonio de Oliveira Salazar
21. Francisco Franco
22. Adolf Hitler
23. Hermann Göring (goe-ring)
24. Joseph Goebbels (goe-bels)
25. Joseph Stalin
26. Pablo Picasso
27. T. S. Eliot
28. John Maynard Keynes (kānz)

a. Physicist who developed theory of relativity (1879-1955)
b. German dictator (1889-1945)
c. Bolshevik leader who organized Red Army (1877-1940)
d. Founder of countercyclical economic theory (1883-1946)
e. Writer who focused on the torture of anxiety (1883-1924)
f. German sociologist who utilized ideal types (1864-1920)
g. Spanish dictator (1892-1975)
h. Leader of Bolshevik revolution and state (1870-1924)
i. Physicist who developed quantum theory (1858-1947)
j. French sociologist who pioneered statistics (1858-1917)
k. Writer from a woman's point of view (1882-1941)
l. Nazi minister-president of Prussia (1893-1946)
m. Psychoanalyst of collective unconscious (1875-1961)
n. Writer with stream of consciousness style (1882-1941)
o. Nazi minister of propaganda (1897-1945)
p. Painter and sculptor in many modern styles (1881-1973)
q. Religious American-British poet (1888-1965)
r. A founder of analytic philosophy (1872-1970)
s. Socialist leader of Russian provisional government (1881-1970)
t. Predictor of *The Revolt of the Masses* (1883-1955)
u. Soviet dictator (1879-1953)
v. Philosopher who focused on symbolic logic (1889-1951)
w. Portuguese dictator (1889-1970)
x. Philosopher of the *Decline of the West* (1880-1936)
y. Physicist who created uncertainty principle (1901-1976)
z. Writer of model interior monologue (1871-1922)
aa. Founder of psychoanalysis (1856-1939)
bb. Italian dictator (1883-1945)

CHRONOLOGICAL DIAGRAMS

Diagram 1

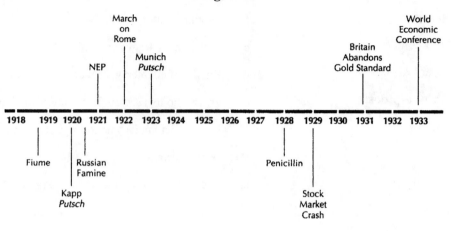

Diagram 2

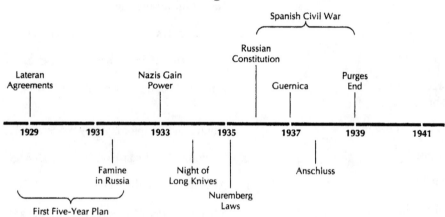

IDENTIFICATION

1. psychoanalysis
2. Surrealism
3. theory of relativity
4. uncertainty principle
5. Bauhaus
6. February Revolution
7. October Revolution
8. War Communism
9. New Economic Policy (NEP)
10. Spartacists
11. Mensheviks
12. Bolsheviks
13. Czech brigade
14. March on Rome
15. fascism
16. totalitarianism
17. Falange
18. *Mein Kampf* (mīn kahmpf)
19. corporative state
20. the SS
21. *Krystalnacht*
22. Nuremberg Laws
23. kulaks

a. Uprising that overthrew the Provisional Government
b. "Minority" faction of Russian Social Democratic Party
c. "Majority" faction of Russian Social Democratic Party
d. Hitler's bodyguards
e. Fascist demonstration that led them to power
f. Artistic style involving dream-like distortions of reality
g. Hitler's book setting out his "philosophy"
h. Bolshevik emergency policy of tight control over economy
i. Expedient Bolshevik policy of allowing private businesses
j. Spanish fascist party
k. Principles relating space and time to velocity
l. Government based on occupational groups
m. Night of violence aimed at Jews in Germany
n. Government system emphasizing discipline and force
o. Center of German modernist movement in art and design
p. Prisoners of war who took over Trans-Siberian railroad
q. Law that there is a physical limit to what can be known
r. Well-to-do peasants in Russia
s. System of government maximizing control over people
t. Statues that stripped German Jews of their rights
u. The study and treatment of people's unconscious minds
v. Uprising that overthrew the Tsar
w. German revolutionary movement at end of World War I

MAP EXERCISE

1. Indicate and label those areas that turned to authoritarianism between 1920 and 1939. Distinguish fascist, communist, and milder authoritarian countries.

PROBLEMS FOR ANALYSIS

I. Two Successful Revolutions

1. Compare the February and the October Revolutions in Russia.
2. Trace how War Communism, the New Economic Policy, and various governmental policies helped the Bolsheviks secure their power and gain some stability in Russia.
3. Analyze the rise to power of the Fascists under Mussolini.

II. The Distinctive Culture of the Twentieth Century

1. Why did Freud's ideas create so much controversy and opposition? In what ways did his ideas relate to other cultural currents of his time?
2. In what ways did the culture of the twentieth century differ from that of the nineteenth century?

III. Totalitarianism

1. What ideas and policies did fascism stand for? What ideas and policies did fascism stand against? What groups found fascism most appealing?
2. What were the principal characteristics of totalitarianism?
3. How do you explain the rise to power of Nazism under Hitler and the great control acquired by the Nazi state?
4. Explain how Stalin managed to preside over rapid industrialization and collectivization and yet retain absolute power.

SPECULATIONS

1. How do you think the Western democracies should have dealt with communism in Russia and fascism in Italy during the 1920s? Why?
2. Suppose you were Lenin or Trotsky; how would you justify the policies pursued by the Russian government between 1919 and 1924?
3. In retrospect, what might have been done within Germany or by forces outside of Germany to prevent the acquisition of power by the Nazis?

TRANSITIONS

In "World War and Democracy," World War I and the period just after that war were analyzed.

In "The Great Twentieth-Century Crisis," the rise of communism in Russia and fascism in Italy is traced. This is followed by an analysis of the retreat of democracy in the face of problems in the 1920s and 1930s, particularly the Great Depression, and the development of totalitarian regimes in Germany under Hitler and in the Soviet Union under Stalin. Some of these developments are shown to have echoes in cultural developments of the early decades of the twentieth century.

In "The Nightmare: World War II," the Second World War and the reconstruction of Europe during the postwar period will be examined.

ANSWERS

Self Test

1d; 2b; 3a; 4d; 5d; 6c; 7d; 8d; 9a; 10b; 11c; 12d; 13d; 14c; 15a; 16d; 17c; 18a; 19a; 20a

Guide to Documents

I-1c; II-1c; III-1a; IV-1b; V-1a

Significant Individuals

1aa; 2m; 3j; 4f; 5z; 6e; 7n; 8x; 9t; 10r; 11v; 12k; 13a; 14i; 15y; 16s; 17h; 18c; 19bb; 20w; 21g; 22b; 23l; 24o; 25u; 26p; 27q; 28d

Identification

1u; 2f; 3k; 4q; 5o; 6v; 7a; 8h; 9i; 10w; 11b; 12c; 13p; 14e; 15n; 16s; 17j; 18g; 19l; 20d; 21m; 22t; 23r

TWENTY-NINE
THE NIGHTMARE: WORLD WAR II

CHAPTER HIGHLIGHTS

1. Democracies, unable to deal effectively with the Great Depression and anti-war sentiments, responded weakly to totalitarian threats, but appeasement led to further demands and finally war.
2. Until the end of 1941, the Axis succeeded—extending its control over much of Europe. From 1942 on, the Allies—by rapidly mobilizing their war effort, holding at Stalingrad, and taking the offense in the air over Europe and on land in North Africa—turned the tide; in 1945 Germany and Japan surrendered.
3. Europe was devastated by the war and was now dependent on the two new superpowers—the United States and the Soviet Union. By 1948 the Cold War had broken out between the two, with the various European nations aligning under the influence and aid of one or the other.
4. In the postwar years Europe recovered quickly. In the East, communist puppet regimes were established almost everywhere; in the West, parliamentary regimes reestablished themselves.
5. Europe gradually withdrew from its colonial empires, sometimes peacefully and sometimes only after bitter fighting. However, even after decolonization, the former colonial powers retained strong commercial, significant cultural, and selective political ties to many former colonies.
6. During the 1950s European nations asserted increased independence from the superpowers within the context of an increasingly stable balance of power between their two blocs.

CHAPTER OUTLINE

I. The Years of Axis Victory
1. The Path to War
2. The Course of the War, 1939-1941

II. The Global War, 1942-1945
1. The Turn of the Tide
2. Competing Political Systems
3. Allied Strategy
4. The Road to Victory

III. Building on the Ruins
1. Immediate Crises
2. The Division between East and West
3. Decolonization

IV. European Recovery
1. Economic Growth
2. New Political Directions
3. The International Context

SELF TEST

1. Hitler's peaceful diplomatic successes in the two years leading up to the war included
   a. the Anschluss, or annexation, of Austria, which had long been a goal of German nationalists.
   b. the annexation of the Sudetenland, the ethnically German border region of Czechoslovakia.
   c. the occupation of Czechoslovakia and part of Lithuania, despite the fact that neither had German ethnics.
   d. the reabsorption of Danzig, the German city that had been made independent to give Poland a port.

2. Britain and France tried to appease rather than resist Germany for all of the following reasons EXCEPT
   a. they had no way of knowing that Hitler glorified war and hoped to create a new German empire.
   b. they hoped that if German's "legitimate" complaints were satisfied, Hitler would be satisfied.
   c. they feared another war, remembering the carnage of World War I.
   d. they needed time to rearm, in case worse came to worst.

3. Stalin concluded an alliance with Hitler because
   a. he was an unscrupulous dictator.
   b. Britain and France had put him off.
   c. he was a crafty Communist.
   d. it was part of his long-term plan to conquer the world.

4. In the first two years of war, Germany overran all of the following EXCEPT
   a. Poland.
   b. Denmark and Norway.
   c. Holland, Belgium, and France.
   d. Yugoslavia, Greece, Crete, and the Soviet Union.

5. The German victories were based on "blitzkrieg" tactics, which involved
   a. a terrible new Teutonic force.
   b. the use of tanks supported by aircraft to drive deep behind enemy lines.
   c. a totalitarian achievement other societies could not equal.
   d. the use of futuristic "lightening" machines that panicked enemy soldiers.

6. The Battle of Britain was important for all of the following reasons EXCEPT
    a. it was the first battle fought entirely between air forces.
    b. it showed that nations were incredibly vulnerable to air attack.
    c. it preserved England as a base for an eventual counteroffensive.
    d. it encouraged Hitler to move east, where he met his doom.

7. The tide of war turned for all of the following reasons EXCEPT
    a. the Japanese attack on the United States brought the world's strongest economy into the allied coalition.
    b. the Russian victory at Stalingrad cost Germany over 300,000 troops, a loss the they could not afford.
    c. the British and Americans quickly mounted an attack into northern France to help the Russians.
    d. British and American victories in North Africa started these allied powers on the road to victory.

8. All of the following states effectively mobilized their people and economies for total war EXCEPT
    a. Germany.
    b. Britain.
    c. the United States.
    d. the Soviet Union.

9. At the height of "production" the death factory at the concentration camp Auschwitz was killing
    a. 1,200 people per week.
    b. 12,000 people per week.
    c. 1,200 people per day.
    d. 12,000 people PER DAY.

10. All of the following were divisive issues in forging Allied strategy EXCEPT
    a. the "Europe first" policy.
    b. when to open the "second front."
    c. who would have how much say in creating new governments like Italy and Poland.
    d. the relationship between the Allies, de Gaulle's Free French, and Vichy officials.

11. The Allied counter-attack included all of the following offensives EXCEPT
    a. a sustained, two-year set of attacks by the Soviets along the breadth of the Eastern front.
    b. Anglo-American attacks on Sicily and then Italy, which brought down Mussolini's government.
    c. an Anglo-American amphibious assault across the English Channel and then to Paris and the Rhine.
    d. an Anglo-Russian offensive in the Balkans that drove the Germans up the Danube valley into Austria.

12. The Yalta conference in February 1945 successfully worked out all of the following EXCEPT
    a. the decision to create and the basic structure of the United Nations.
    b. the way governments would be set up in liberated countries.
    c. the entry of the Soviet Union into the war against Japan.
    d. the status of France in the occupation of Germany.

13. The devastation of the war in Europe included all of the following EXCEPT
    a. the war killed approximately 50,000,000 people (!).
    b. Europe's industrial production was half (1/2) of what it had been in 1939.
    c. 60,000,000 (sixty MILLION) Europeans were refugees at the end of the war.
    d. Atom bombs had wiped out entire cities, leaving them heaps of rubble.

14. All of the following were imposed on Germany EXCEPT
    a. it was divided into four occupation zones.
    b. it was saddled with a huge reparations debt.
    c. its leaders were put on trial for war crimes.
    d. its eastern border was adjusted to the west.

15. The Allies created all of the following international agencies to help maintain peace EXCEPT
    a. the United Nations Organization, to provide relief and, if necessary, keep the peace with armed force.
    b. Amnesty International, to keep track of the human rights abuses that had characterized the Axis.
    c. the International Monetary Fund, to keep currencies stable.
    d. the International Bank for Reconstruction and Development, which later became the World Bank.

16. The new regimes were characterized in all of the following ways EXCEPT
    a. Western European countries re-established parliamentary regimes.
    b. Eastern European countries got governments and economies modeled on the Soviet Union.
    c. Germany re-established the Weimar Republic based on its old constitution.
    d. England went through a profound transformation under a socialist Labor government.

17. The early Cold War involved all of the following events in Europe EXCEPT
    a. the Berlin airlift.
    b. the Greek rebellion.
    c. the Marshall plan.
    d. the Korean War.

18. Early colonial losses after World War II included all of the following EXCEPT
    a. Britain gave up control of India and Pakistan.
    b. Holland lost control of Indonesia.
    c. France was defeated by rebels in Vietnam and Algeria.
    d. Germany lost Togo and Cameroon.

19. Decolonization had the effect of
    a. replacing European with Soviet and American control.
    b. replacing direct control with indirect influence.
    c. ending outside interference in the newly established states.
    d. ending internal and external conflicts involving the former colonies.

20. In 1952, the combined Gross National Product (GNP) of the OEEC countries was

    a. 50 percent of 1938.

    b. equal to 1938.

    c. one and one half times 1938.

    d. double 1938.

21. All of the following countries experienced stability in the 1950s EXCEPT

    a. Britain, which after the war had put into place the most comprehensive social welfare system.

    b. West Germany, which under Adenauer established itself as a pillar of the Western community.

    c. France, which came under the influence of de Gaulle once again toward the end of the decade.

    d. the Soviet Union, which experienced a relaxation under Khrushchev after Stalin died.

22. Europeans established greater independence in all of the following cases EXCEPT

    a. Britain and France invaded Egypt with Israel despite American and Soviet disapproval.

    b. East Germany, Romania, and Poland gained greater autonomy from Russia.

    c. Albanian aligned itself with Communist China rather than the Soviet Union.

    d. Yugoslavia played the United States and the Soviet Union off against each other.

GUIDE TO DOCUMENTS

I. Stalin Appeals to Patriotism

1. How does Stalin appeal to citizens for support?

    a. As an extension of class war between capitalists and the proletariat.

    b. As an international struggle for freedom and against fascist enslavement.

    c. As a matter of state in which the people must do their duty.

    d. As a battle to the death between the Soviet and the German people.

2. What strategy does Stalin pursue in the face of Nazi victories?

II. A Crematorium

1. What methods were used to effect these massive killings?

    a. The victims were drowned.

    b. The victims were shot.

    c. The victims were gassed.

    d. The victims were burned.

2. What is the significance of what was done to the bodies after the killings?

III. The Historians' Debate on German Genocide

1. What is Friedlander's point?

    a. The crucial issue of the Holocaust is to determine exactly who gave the order and when it was given.

    b. It will be very difficult to determine exactly when Hitler personally shot Jews as part of the Holocaust.

    c. The decision to exterminate the Jews was a direct outgrowth of racist German genetic "science."

    d. It will never be possible to determine exactly who was responsible for the Holocaust or why it happened.

2. The implication of Nolte's analysis is that

    a. the Holocaust was not primarily a result of traditional anti-Semitism and was not mere genocide.

    b. the Nazis' fear of the Jews was a distorted reaction to the Communists atrocities in Russia.

    c. the Nazis were worse than the Communists because they were more cold-blooded and efficient.

    d. the Communists are also to blame for the Holocaust since the Nazis were following their lead.

3. The strongest piece of evidence Wehler gives for his overall argument is that

    a. Hitler was a rabid anti-Marxist long before the Russian Revolution.

    b. Hitler was strongly influenced by Social Darwinism long before the Russian Revolution.

    c. a direct call for the extermination of the Jews is on record from 20 years before the Russian Revolution.

    d. the Nazis adopted long-standing general reactionary attitudes from their environment in Germany.

4. Which historian do you agree with? Does it matter which one is right? Why or why not?

IV. Churchill Sees an Iron Curtain

1. What was meant by Churchill's statement that "an iron curtain has descended across the Continent"?

    a. Europe was being divided by Soviet political domination of areas it occupied into two hostile camps.

    b. The Soviets were constructing the physical barrier that would later be called the "Berlin Wall."

    c. The Soviet army, with its tanks and artillery, was cutting Europe in two, sealing off the Soviet zone.

    d. The Soviets were making a shield the Germans could hide behind to protect themselves from the Allies.

2. What policies might flow from Churchill's perceptions as revealed in this excerpt?

V. The Soviet Union Denounces the United States While Calling for Arms Reduction

1. Vishinsky cites all of the following evidence to support his views EXCEPT

    a. American support for conservative regimes against national liberation movements.

    b. American organization of military alliances and activation of military bases.

    c. American use of the atomic bomb in an attempt to intimidate the Soviet Union.

    d. American propaganda promoting the idea that the Soviet Union is planning a war.

2. How do Vishinsky's views differ from those of Churchill?

SIGNIFICANT INDIVIDUALS

1. Neville Chamberlain
2. Charles de Gaulle (gal)
3. Pierre Laval
4. Winston Churchill
5. Franklin Roosevelt
6. Dwight Eisenhower
7. Harry Truman
8. Marshal Tito
9. Eduard Benes (be-nesh)
10. Konrad Adenauer
11. Mohandas Gandhi
12. Nikita Khrushchev (KROOSH-tchoff)
13. John Foster Dulles
14. Gamal Abdel Nasser
15. Walter Ulbricht
16. Wladyslaw Gomulka

a. Bulldog who lead Britain in its darkest hour (1874-1965)
b. Independent Yugoslav Communist leader (1892-1980)
c. Non-violent leader of Indian independence (1869-1948)
d. Commander of Anglo-American forces (1890-1969)
e. American president at end of war (r.1945-1953)
f. Appeasing British Prime Minister (1869-1940)
g. Flamboyant Soviet leader after Stalin (1894-1971)
h. American president and Allied leader (r.1933-1945)
i. America secretary of state in 1950s (r.1953-1959)
j. Czech leader pushed out by Soviets (1884-1948)
k. Nationalist Polish Communist leader (1905-1982)
l. Nationalist Egyptian leader (1918-1970)
m. First leader of postwar West Germany (r.1949-1963)
n. Leader of Free French and Fifth Republic (1890-1970)
o. Vichy French leader (1883-1945)
p. Communist leader of East Germany (1893-1973)

CHRONOLOGICAL DIAGRAMS

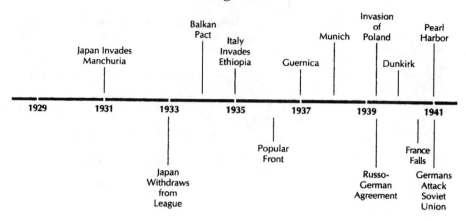

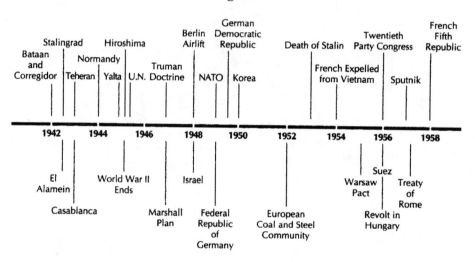

IDENTIFICATION

1.	Anschluss (AHN-schloos)	a.	Nazi program to kill all the Jews
2.	Sudetenland (soo-DĀT-n-land)	b.	Lightning war utilizing planes, tanks, and trucks
3.	blitzkrieg (BLITS-krēg)	c.	City obliterated by first atomic bomb
4.	Ultra project	d.	European economic structure to funnel Marshal Plan aid
5.	El Alamein (al-a-MĀN)	e.	Site and name of last wartime Allied summit meeting
6.	Stalingrad	f.	Military alliance of Soviet Union and Eastern Europe
7.	"the final solution"	g.	American economic program to help rebuild Europe
8.	partisans	h.	Allied code-breaking effort
9.	second front	i.	Prosecutions of German leaders for their crimes
10.	Yalta	j.	The German annexation of Austria
11.	Hiroshima	k.	The main American and British thrust against Germany
12.	Nuremberg trials	l.	Soviet meeting where Khrushchev bared Stalin's crimes
13.	Berlin airlift	m.	Western Allies' crucial victory in North Africa
14.	Cold War	n.	Section of Czechoslovakia containing ethnic Germans
15.	Truman Doctrine	o.	European trading block
16.	Marshall Plan	p.	Civilians carrying on a guerrilla war behind the lines
17.	NATO	q.	Global struggle between the US and the Soviet Union
18.	Warsaw Pact	r.	Use of aircraft to defeat Soviet siege of German capital
19.	OEEC	s.	City where Soviets turned the tide of war against Nazis
20.	Twentieth Party Congress	t.	Military alliance of United States and Western Europe
21.	European Economic Community	u.	American strategy for resisting expansion of Soviet power

MAP EXERCISES

1. Above is a map of Europe in 1939 before the outbreak of World War II. Indicate the political and territorial changes that would result in 1945 after the war.

1. Label the political divisions of Europe, and indicate the spheres of influence of the United States and the Soviet Union as of about 1950.
2. Indicate those nations that formed the European Economic Community in 1957.

PROBLEMS FOR ANALYSIS

I. The Years of Axis Victory

1. Evaluate how effectively the democracies responded to the growth of authoritarianism and to such problems as the Ethiopian crisis, the Anschluss, and the Czech partition.
2. France was one of the dominant powers in Europe, yet it fell relatively quickly to the Germans in 1940. How do you explain this collapse?

II. The Global War, 1942-1945

1. Analyze the reasons for the turning of the tide against the Axis in 1942. What factors were weakening the Axis powers? What factors were strengthening the Allied powers?

III. Building on the Ruins

1. Compare the peace settlement after World War II with that after World War I. Do you think that the Allied leaders learned from history? Were the lessons valid?
2. Compare the political evolution of Eastern and Western Europe between 1945 and 1950.
3. Trace the evolution of the Cold War in the late 1940s and early 1950s. Why was competition between the two great powers so intense?

IV. European Recovery

1. What roles did the United States and the Soviet Union play in the economic recovery of Europe? How important was the Marshall Plan? Did the United States and the Soviet Union benefit by providing recovery aid? Explain.
2. Do trends between 1945 and 1958 support hope for the development of regionalism or of internationalism? Explain.
3. Evaluate the attempts by various European states to achieve increased independence from US or Soviet influence.

SPECULATIONS

1. In retrospect, what policies might have succeeded in stopping Hitler and Mussolini and avoiding World War II? What prevented statesmen or nations from following such policies?
2. As a French communist supporter of the Soviet Union in 1939, how would you explain the Soviet-German agreement? What arguments would you be likely to encounter?
3. Imagine that you are in the crematorium at Birkenau with members of your family. Describe the 10 minutes between the time the doors are forced shut and when the gas kills you.
4. Assume you are a Soviet statesman during the late 1950s. Defend the Soviet Union's policies in Europe between 1945 and 1955, and attack those of the United States.
5. Assume you are an American statesman during the late 1950s. Defend America's policies in Europe between 1945 and 1955, and attack those of the USSR
6. Considering the devastation caused by World War II, the loss of colonial empires following the war, and the rise to superpower status of the United States and the USSR, should the mid-twentieth century be considered the culmination of Europe's decline? Explain.

TRANSITIONS

In "The Great Twentieth-Century Crisis," the growth of communism and fascism, the development of totalitarianism, and the culture of the twentieth century were examined.

In "The Nightmare: World War II," the retreat of democracies in the 1930s and the years of World War II are traced. Ultimately the Allies were victorious, but after the war European nations faced devastation, dislocation, loss of position in the world, and possible domination by the two new competing superpowers: the United States and the Soviet Union. Yet Europe recovered remarkably well, creating growing societies relatively consistent with traditional liberal ideals.

In "Contemporary Europe," recent decades will be analyzed, and some of the crises and problems facing us in today's world will be evaluated.

ANSWERS

Self Test

1d; 2a; 3b; 4d; 5b; 6b; 7c; 8a; 9d; 10a; 11d; 12b; 13d; 14b; 15b; 16c; 17d; 18d; 19b; 20c; 21c; 22a

Guide to Documents

I-1b; II-1b; III-1c; III-2d; III-3c; IV-1a; V-1c

Significant Individuals

1f; 2n; 3o; 4a; 5h; 6d; 7e; 8b; 9j; 10m; 11c; 12g; 13i; 14l; 15p; 16k

Identification

1j; 2n; 3b; 4h; 5m; 6s; 7a; 8p; 9k; 10e; 11c; 12i; 13r; 14q; 15u; 16g; 17t; 18f; 19d; 20l; 21o

THIRTY
CONTEMPORARY EUROPE

CHAPTER HIGHLIGHTS

1. Europe has made substantial moves toward economic integration, but its failure to react effectively to the savage turmoil accompanying the breakup of Yugoslavia demonstrates the depth of its political divisions.
2. Europe's postindustrial society remains wealthy and advanced. Increased independence from the superpowers, movements toward unity, and searches for political alternatives evidence continued European vitality.
3. Eastern Europe experienced upheaval and transformation with the fall of communism. Western Europe has maintained a fundamental stability of democratic institutions, though problems of economic dislocation, energy shortage, and terrorism continue to frustrate Europeans.
4. The fall of Communist regimes in Eastern Europe and the Soviet Union marked the end of the Cold War and the end of an era.
5. A burst of creativity and new ideas such as postmodernism have marked cultural and intellectual life in recent decades.

CHAPTER OUTLINE

I. The New Institutions
1. Cautious Beginnings
2. Toward European Union

II. Postindustrial Society
1. Europe's Advantage

III. The Politics of Prosperity
1. Waves of Protest
2. Capitalist Countries: The Challenge of Recession
3. Communist Rule: The Problem of Rigidity

IV. The End of an Era
1. The Miracles of 1989
2. The Disintegration of the USSR
3. Europe without Cold War

V. Contemporary Culture
1. Postwar Creativity
2. The Explosion of Popular Culture
3. Social Thought

SELF TEST

1. All of the following contributed to the concept of European unity EXCEPT
   a. Germany's success during World War II in integrating the areas under its rule in a unified empire.
   a. Allied propaganda emphasizing the shared values of Western civilization during World War II.
   b. the success of transnational programs in fostering the postwar economic recovery.
   c. Cold War military alliances.

2. All of the following are strong examples of the progress of European integration EXCEPT
   a. the formation of the Council of Europe in 1948.
   b. the creation of the European Economic Community in 1957.
   c. the expansion of the EC both in membership and in the scope of its activities.
   d. the reaction to the Maastricht Treaty of 1992 transforming the EC into the European Union.

3. The strongest evidence of the European Union's continuing vitality is
   a. the votes on acceptance of the Maastricht treaty.
   b. the reaction of the members to the common currency.
   c. the number of countries that are applying for membership.
   d. its declaration of human rights.

4. All of the following contribute to Europe's continuing economic strength EXCEPT
   a. its advanced transportation infrastructure.
   b. its enthusiastic investments in technology.
   c. the addition of the former East Bloc's manufacturing plant.
   d. its high levels of schooling and strong educational programs.

5. Women's lives have changed in all the following ways EXCEPT
   a. their participation in education almost equals men's.
   b. their pay has come to be almost equal to men's.
   c. labor unions have begun to champion their interests.
   d. contraception and abortion are readily available.

6. The major socioeconomic challenges facing Europe include all of the following EXCEPT
   a. maintaining the social welfare systems in the face of recession and endemic unemployment.
   b. peacefully integrating the immigrants from less developed areas of the world.
   c. stemming the outflow of technically skilled workers to the United States.
   d. protecting the environment despite the cost to already hard-pressed economies.

7. The radicals in the late 1960s failed to spark widespread revolutions for all the following reasons EXCEPT
    a. labor unions and the traditional left were suspicious of them.
    b. liberals and conservatives were put off by their manners and tactics.
    c. the voters repudiated them at the polls.
    d. the security forces massacred them.

8. Women in the student movement turned to feminism because they became disillusioned by
    a. the unequal treatment they got from male radicals.
    b. violence of the radical fringe.
    c. the hostility they faced from ordinary citizens.
    d. the futility of the attempt to spark political revolution.

9. European terrorists used all of the following tactics EXCEPT
    a. kidnapping.
    b. assassination.
    c. poison gas.
    d. bombings.

10. The European recession was caused by all of the following EXCEPT
    a. radical sabotage.
    b. the energy crisis.
    c. inflation.
    d. economic stagnation.

11. The Soviet Union's failure to keep up economically was evidenced by all of the following EXCEPT
    a. its lagging agricultural productivity.
    b. its failure to develop high tech industries.
    c. the performance of its heavy industry.
    d. its weakness in providing services.

12. Gorbachev's reform efforts focused on all of the following EXCEPT
    a. *perestroika*, or political and economic restructuring.
    b. *glasnost*, or openness in public discussion and debate.
    c. the unstated but subtle dismantling of communist party influence over the country.
    d. reduction of international tensions in order to free up resources from the military.

13. The major challenge to Gorbachev's reforms came from
    a. disgruntled party members.
    b. the rise of nationalism.
    c. traditional dissidents.
    d. the re-emergence of class conflict.

14. Communist governments fell in Eastern Europe because of all of the following EXCEPT
    a. they were unable to solve chronic economic problems.
    b. NATO threatened war if the Soviets moved against reformers.
    c. Gorbachev declined to intervene to prop them up.
    d. the reform movements commanded widespread support.

15. Boris Yeltsin replaced Gorbachev as the commanding figure when
    a. he took the initiative to intervene on behalf of communist regimes in Eastern Europe.
    b. he stood up to the old-line Communists who tried to stage a coup while Gorbachev was away.
    c. he led a coup that seized power from Gorbachev and took over the Soviet government himself.
    d. Gorbachev was killed in the abortive coup by radical reformers, who Yeltsin then suppressed.

16. The Soviet Union ceased to exist when
    a. the Baltic Republics broke away.
    b. the leaders of Russia, Ukraine, Belarus, and Kazakhstan said so.
    c. Boris Yeltsin suppressed the Communist-dominated parliament.
    d. the army refused to intervene in breakaway republics.

17. The end of the Cold War brought all of the following changes EXCEPT
    a. the return of Moldova to Romania.
    b. the reunification of Germany.
    c. the bloody disintegration of Yugoslavia.
    d. the separation of the Czech Republic and Slovakia.

18. The disintegration of Yugoslavia included all of the following EXCEPT
    a. the secession of Croatia, Slovenia, Macedonia, and Bosnia-Herzegovina.
    b. a brutal civil war in Bosnia between ethnic Serbs and Muslims.
    c. the separation of the Czech Republic and Slovakia.
    d. a secessionist movement in Kosovo that led to a massive air campaign against Serbia by NATO.

19. Europe's response to the violence accompanying the disintegration of Yugoslavia can best be characterized as
    a. vigorous.
    b. ineffectual.
    c. principled.
    d. reckless.

20. Europe's cultural renewal included all of the following EXCEPT
    a. existentialism, which said that even without moral certainty, a person can still choose to live a moral life.
    b. Vatican II, which modernized Church ritual and beliefs.
    c. the revival of classical literary forms in an attempt to reach back to the roots of European culture.
    d. artistic movements that picked up and carried forward earlier modernist themes and techniques.

21. Popular culture showed all of the following EXCEPT
    a. the strong influence of America.
    b. a strong orientation toward youth.
    c. a surprising dedication to austerity and form.
    d. a desire to experiment and a readiness to shock.

22. Contemporary social thought has come to focus on
    a. the purposes of the biases that structure thought itself.
    b. the delineation of the external forces that determine human fate.
    c. a recognition of the ultimate futility of intellectual endeavor.
    d. a renewed commitment to the core values of Western Civilization.

GUIDE TO DOCUMENTS

I. A Turkish Girl Arrives in Germany

1. According to Aynur's account, all of the following tensions were present in her experience EXCEPT?
    a. the difference between her expectations about what their housing would be like and what it was.
    b. the difference between her living space in Turkey and those in Germany.
    c. the difference between appropriate dress for a girl in Turkey and appropriate dress in Germany.
    d. the difference between her father's response to Germany's standards of female dress and her own.

2. Using this excerpt as a source, how would you evaluate the experience of immigrating from Turkey to Germany?

II. Havel's Inaugural Address

1. Havel thinks that the worst legacy of communism for Czechoslovakia was
    a. the inefficiency of the economy.
    b. the pollution of the environment.
    c. the corruption of human relations.
    d. the misuse of agricultural resources.

2. How does the substance of Havel's speech embody the remedy that he thinks is needed for the country's ills?

III. Foucault on Sexual Discourse

1. According to Foucault, all of the following relationships are true EXCEPT
    a. repression of sex is encouraged by the liberating thrill of defiance people then get simply talking about it.
    b. the association of defiance and sex makes revolution essentially an act of sensual gratification.
    c. the linking of sex and revolution (expectation of what will be) makes talk about sex a form of prophesy.
    d. the relation of sex, change, prophecy, and attaining bliss makes sex talk the modern form of preaching.

2. In what ways does sex serve as a support for preaching? Why, according to Foucault, is this so important?

IV. A Reflection on Contemporary Feminism

1. According to Kristeva, phase one of the women's movement focused on specific legal and political issues, and

    a. phase two will focus on issues of time and place.

    b. phase two will gain women a place in linear time as the time of project and history.

    c. phase two is creating a language for women to express universals that have been heretofore inexpressible.

    d. phase two will cultivate the role of the woman as mother.

2. What does the passage imply about why the women's movement has moved into a new phase? How is this new phase related to the desire of more active women to be mothers? How is it related to religion?

SIGNIFICANT INDIVIDUALS

1.	Willy Brandt	a.	Leader of the collapse of the Soviet Union (1931-)
2.	Helmut Kohl	b.	Czech leader after end of Communism (1936-)
3.	Margaret Thatcher	c.	Liberalizing pope who called Vatican II (r.1958-1963)
4.	Lech Walesa (lek vuh-lenss-uh)	d.	American president who sold Soviets wheat (1913-1994)
5.	Alexander Dubcek (dub-chek)	e.	Russian author of *Dr. Zhivago* (1890-1960)
6.	Vaclav Havel	f.	Czech leader during the "Prague Spring" (1921-)
7.	Nicolae Ceausescu (sho-SHESS-coo)	g.	American president who helped spend the Soviet Union into the ground (1911-)
8.	Leonid Brezhnev (BREZH-neff)	h.	Prime Minister who recast British economy (1925-)
9.	Mikhail Gorbachev	i.	Chancellor who reunified Germany (1930-)
10.	Richard Nixon	j.	Leading poststructuralist philosopher
11.	Ronald Reagan	k.	French philosopher of cultural history
12.	John XXIII	l.	French existentialist author (1913-1960)
13.	John Paul II	m.	Polish pope shot by a Turkish terrorist (r.1978-)
14.	Boris Pasternak	n.	Anthropologist and structuralist (1908-)
15.	Albert Camus	o.	Tyrannical Communist boss of Romania (1918-1989)
16.	Claude Lévi-Strauss	p.	Soviet leader from the 1960s to the 1980s (1907-1982)
17.	Alexander Solzhenitsyn (zol-je-NĒT-sin)	q.	West German chancellor who fostered improved relations with the Soviet Bloc (1913-1992)
18.	Jean Paul Sartre	r.	French existentialist philosopher (1905-1980)
19.	Michel Foucault (foo-KŌ)	s.	Leader of Solidarity trade union (1943-)
20.	Jacques Derrida	t.	Russian author who exposed the gulag (1918-)

CHRONOLOGICAL DIAGRAM

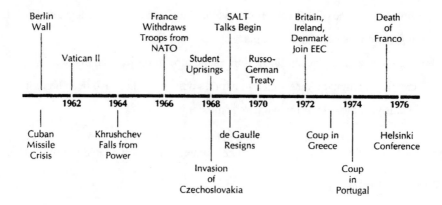

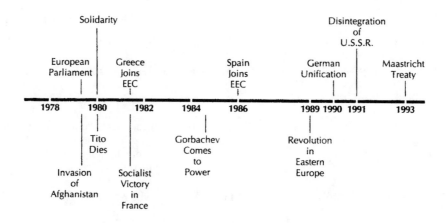

IDENTIFICATION

1. "opening to the left"
2. Irish Republican Army (IRA)
3. Red Brigades
4. Eurocommunism
5. Brezhnev doctrine
6. Council of Europe
7. stagflation
8. Solidarity
9. détente
10. European Economic Community
11. postindustrial society
12. existentialism
13. structuralism
14. *perestroika*
15. postmodernism
16. deconstruction
17. Confederation of Independent States
18. *Parc Asterix*

a. Most successful aspect of European integration
b. Early loose European union
c. Competitor to Eurodisney
d. Independent trade union in Poland
e. Policy of political and economic restructuring
f. Loose affiliation of current ideas
g. Attempt to woo Italian socialists from communists
h. Loose union that followed Soviet Union
i. Policy that threat to any communist regime threatened all
j. Philosophy that meaning is what we make it
k. Philosophy that meaning is what it makes us
l. Technique of analyzing texts to find their many meanings
m. Militant nationalists on the Emerald Isle
n. Economic condition of inflation plus slow growth
o. West European communists' attempt to be independent
p. America-led policy of accommodation with Soviet bloc
q. Stage when more people work in services than manufacturing
r. International terrorist organization

MAP EXERCISE

1. Indicate those areas experiencing revolts and major political changes in the period between 1989 and 1994.

PROBLEMS FOR ANALYSIS

I. The New Institutions

1. In what ways has European integration succeeded? In what ways has it failed? How can you account for these successes and failures?

II. Postindustrial Society

1. Does the evidence of the past 30 years support the argument that Europe is suffering from a long-term decline?
2. How have governments attempted to deal with problems accompanying economic growth over the past 30 years?

III. The Politics of Prosperity

1. In what ways did Eastern and Western European nations face different sorts of problems during the 1960s, 1970s, and 1980s?
2. How could there be so many alternations of political power within Western European nations and still so much fundamental political stability?

IV. The End of an Era

1. How do you explain the fall of Communism in 1989 and 1990?
2. What has been the significance of this fall of Communism in recent years?

V. Contemporary Culture

1. What developments indicate that Europe's culture and intellectual life during the past 30 years have remained creative and productive?
2. How do developments such as postmodernism and poststructuralism offer fundamental criticisms of previous intellectual and cultural trends?

SPECULATIONS

1. Suppose the next 30 years were to be characterized by logical extensions of the main developments over the past 30 years. Describe the history of the next 30 years.
2. Considering the past 30 years in the West, in what ways are these the best of times and in what ways are these the worst of times? Explain.
3. In what ways might it be argued that 1989 marks the end of a historical era covering much of the twentieth century? What should be the beginning date of such an era?

TRANSITIONS

In "The Nightmare: World War II," the war and the postwar era were traced, with emphasis on the recovery of Europe in a world dominated by two competing superpowers: the United States and the Soviet Union.

In "Contemporary Europe," developments over the last 30 years are examined. The period is marked by continued economic growth and efforts to deal with problems stemming from that growth as well as a series of new cultural developments such as postmodernism. Most striking and of great potential significance is the fall of Communism in Eastern Europe and the Soviet Union in 1989–1990. It remains unclear how Europeans will deal with these challenges in the coming years.

ANSWERS

Self Test

1a; 2d; 3c; 4c; 5b; 6c; 7d; 8a; 9c; 10a; 11c; 12c; 13b; 14b; 15b; 16b; 17a; 18c; 19b; 20c; 21c; 22a

Guide to Documents

I-1b; II-1c; III-1b; IV-1c

Significant Individuals

1q; 2i; 3h; 4s; 5f; 6b; 7o; 8p; 9a; 10d; 11g; 12c; 13m; 14e; 15l; 16n; 17t; 18r; 19k; 20j

Identification

1g; 2m; 3r; 4o; 5i; 6b; 7n; 8d; 9p; 10a; 11q; 12j; 13k; 14e; 15f; 16l; 17h; 18c

SECTION SUMMARY
THE TWENTIETH CENTURY
1914–PRESENT
CHAPTERS 27–30

CHRONOLOGICAL DIAGRAM

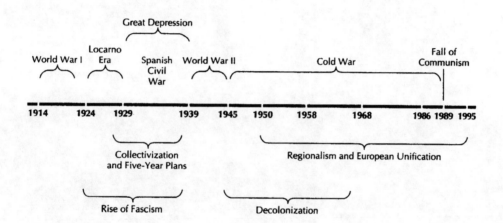

MAP EXERCISES

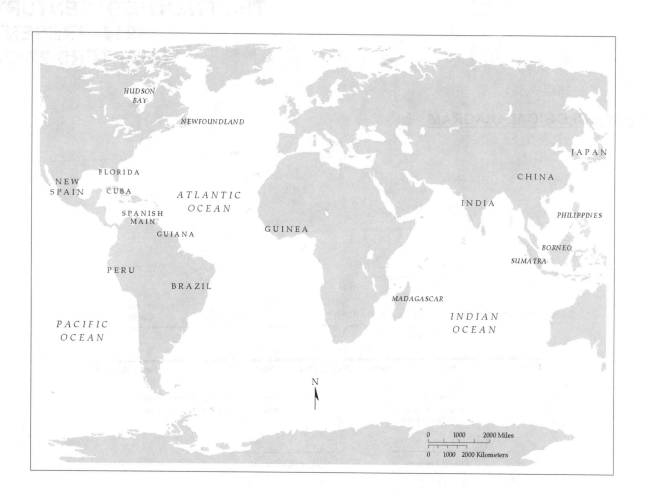

1. Indicate the areas controlled by the imperial powers in 1914.
2. Indicate the approximate dates when these areas gained independence during the period 1945–1965.

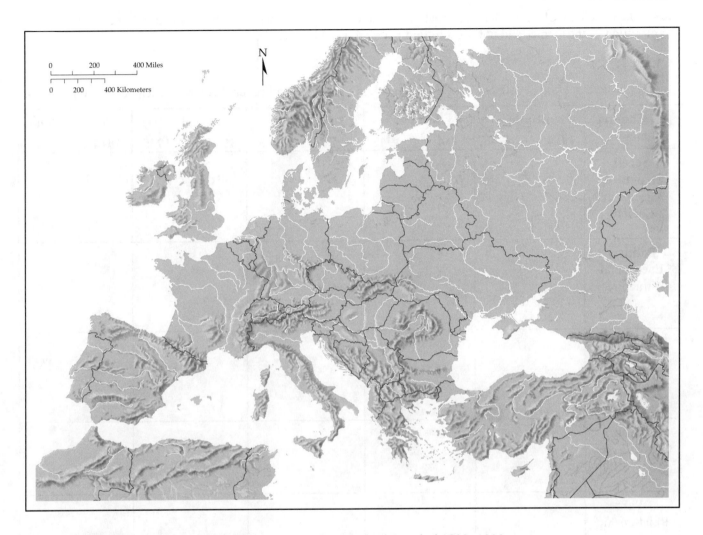

1. Indicate the spread of authoritarianism and fascism in the period 1922–1939.
2. Indicate the American and Russian spheres of influence in 1950.
3. Indicate the countries in the European Community in 1994.

BOX CHARTS

Reproduce the Box Chart in a larger format in your notebook or on a separate sheet of paper. For a fuller explanation of the themes and how best to find material, see Introduction.

Chart 1:

It is suggested that you devote one page for each column (i.e., chart all seven themes for each country on a page).

Country Themes	Britain		France		Germany		Italy		Soviet Union	
	1914	1945	1914	1945	1914	1945	1914	1945	1914	1945
Social Structure: Groups in Society										
Politics: Events and Structures										
Economics: Production and Distribution										
Family Gender Roles Daily Life										
War: Relationship to larger society										
Religion: Beliefs, Communities, Conflicts										
Cultural Expression: Formal and Popular										

Chart 2:

It is suggested that you devote one page for each column (i.e., chart all seven themes for each country on a page).

Country Themes	Britain 1945	Britain Present	France 1945	France Present	Germany 1945	Germany Present	Italy 1945	Italy Present	USSR/Russia 1945	USSR/Russia Present
Social Structure: Groups in Society										
Politics: Events and Structures										
Economics: Production and Distribution										
Family Gender Roles Daily Life										
War: Relationship to larger society										
Religion: Beliefs, Communities, Conflicts										
Cultural Expression: Formal and Popular										

Chart 3:

It is suggested that you devote one page for each row (i.e., chart each of the seven themes on a separate page; you may want to use blank or graph paper and turn the page sideways). Note major national differences, but keep the focus on development in Europe as a whole.

Periods Themes	1910	1925	1940	1970	Present
Social Structure: Groups in Society					
Politics: Events and Structures					
Economics: Production and Distribution					
Family Gender Roles Daily Life					
War: Relationship to larger society					
Religion: Beliefs, Communities, Conflicts					
Cultural Expression: Formal and Popular					

CULTURAL STYLES

1. Compare the pictures on pages 972 and 973 with those on pages 962, 976, and 982. What do they suggest about the difference between people's expectations of war in 1914 and the reality of World War I?

2. Compare the pictures on pages 991 with those on pages 964, 771, and 526. What is similar about these pictures? What is different?

3. Contrast the picture on page 1928 with that on 1029. What do they suggest about the difference between aesthetic values of elite culture and popular cultures at the time?

4. Compare the posters on pages 1019 and 1048. What is similar about them? What is different? Do you think these similarities and differences manifest similarities and differences in the movements they represent?

5. Now compare also the photographs on pages 1013 and 1016. What is similar about them? What is different? Do you think these similarities and differences manifest similarities and differences in the movements they represent?

6. Examine the pictures on pages 1017, 1031, and 1040. What do they suggest about the values of fascism and Nazism? What do they suggest about their appeal?

7. Look at the pictures on pages 1067, 1068, 1088, and 1089. What do they show about the nature of World War II?

8. Examine the pictures on pages 1075 and 1078. What do they show about modern society?

9. How does the picture on page 1082 compare with the ones you looked at in question 2? What do the differences suggest about the changes to Europe's international relations brought about by World War II.

10. Examine the picture on pages 934, 936, 1023, 1051, and 1177. What development do you see in modern art?

11. Look at the picture on page 1131. What does it suggest about the faith of modern leaders in the power of images to shape people's perceptions of reality? What does it suggest about the larger relationship of image and reality in contemporary society?

EPILOGUE

HIGHLIGHTS

1. The historical perspective suggests caution when analyzing developments in the present, such as the changes in Eastern Europe.
2. The Western preoccupation with economic growth has spread around the globe. Europe will not enjoy the same competitive advantage it has had in the past, and may reach limits to growth set by the environment.
3. A reaction against the size and costs of large governments has set in, even as people expect the state to provide unprecedented levels of services.
4. Some Western values are becoming accepted as universal human rights, even as Western culture has begun questioning the universality of its beliefs and celebrating the diversity of mulitculturalism.
5. The content of the social contract has become less clear, while the role of culture in defining community and the relationships within it has become a central object of both theoretical investigation and practical action.
6. We tend to overestimate the importance of the present and romanticize the past. The future is inherently uncertain, but, as the example of the collapse of communism shows, it is important to keep in mind that human choices, both those of the powerful and those of ordinary citizens, can make a profound difference in the way it unfolds.

OUTLINE

I. The Present and the Past
1. Looking for Lessons

II. The Modern Economy
1. National Standing

III. The Functions of the State
1. Economic Policies
2. Diminishing the State

IV. Questions of Values
1. Human Rights

V. The Nature of Community
1. Cohesion and Conflict
2. Splintered Cultures